ASIA PACIFIC INTERIOR DESIGN AWARDS FOR ELITE

可持续舒适空间

亚太室内设计精英邀请赛获奖作品精选集（下）

亚太设计中心 编著

Claude Bérubé（加拿大）王怡（中国）主编

江苏人民出版社

Approximately 15 years ago the British Architect Sir Richard Rogers was looking into the future of Architecture. At that time he was convinced, that the Design of the future will close the ranks of high end techniques and ecological approaches. Today this is not a scenario of the future anymore; we can see the translation of this statement in Architecture and Design as well. The huge competitions show the focus on this issue, nearly each municipality forces it Architects to make this a major topic in their work.

From my point of view, even I agreed with Richard for 100 % at that time, sustainability (how I do hate this word, it is used as an apology, a concept, a vision or justification for bad design) is not a technical issue at first. Primarily it is about creating a society which understands and accepts their new situation in our globalized world. Even the process of Globalization is mainly to do with economics; our environment shows us to take its needs into consideration whatever we do. The 18th and 19th century was the time of the industrial revolution which marked the transition between the agricultural society into the industrial society. In the 20th century it was replaced by a service society in which people started to reach the goals of the "creation of value process" not by producing goods but by service and service related activities. People questioned economical basics not for the sake of socio-economical developments, but for the egoistic approach in a world which got dominated by materially-approaches and the constant desideratum of the achievements of capitalism. This time was peeled away from a society which had to learn that humankind cannot dictate endlessly its environment what to accept and what not. To the contrary, nature started to dictate what is possible or not. This dictation was done with the most extreme consequence human beings can imagine: Without the acceptance there will be, sooner or later, no life anymore on earth. At the same time socialistic and communistic regimes in Europe collapsed and the world had to face other challenges as well. How to make these people understand the necessity of environmental behavior when trying to reach the standard of life these people for decades were looking for? I am sure, the future will be coined by this kind of questions in political and social developments worldwide.

Coming back to Architecture. I discuss quiet often with my students projects which are showcased in glossy magazines. Students are fascinated by form and material. The question for me is not what the building looks about at the first place; my question is to do with the effect of learning when designing the project. Without any doubt, nowadays nearly anything is possible in terms of technique and material. The interesting issue is the holistic approach which is to do with design, technique, tradition, society, culture, politics and so much more. I believe that only the integration of all these structures will lead to a successful and "sustainable" project which deals with the challenges of the 21st century.

Architecture cannot change the world for sure, but it can integrate the questions this world is asking, trying to submit a direction which could lead to some answers.

I believe in the young generation of Designers and Architects. The flow of information is constant. If there is an interesting project finished in China, committed Architects have knowledge about it within seconds. The social networks connect people worldwide and develop a digital space which serves a need which cities tried to satisfy so far; Communication. The size of this digital city is much bigger than the traditional structure. Hence the flow of information and communication is much more intense and high speeded. This leads to knowledge, at the same time to awareness for problems and their solution as well. I consider these techniques as a tool which creates a huge chance for environmental awareness and cultural intersection.

I was surprised to see only a few projects which really try to deal with the topic of the competition. I saw a lot of design, good and bad. Nowadays the focus on design only is not enough anymore, its a dying discipline. The refusal of accepting the "close of the ranks" between the design and technique as due to Rogers is extremely obsolete. But anything needs time, I am convinced in another 15 years no competition brief has to mention the need for "green" results any more.

I wish the organizers a successful outcome of this important event.

Article

评委序言

Michael Schwarz

大约15年前，英国建筑师理查德·罗杰斯爵士曾探寻建筑设计的未来趋势，当时，他确信未来的设计将与高端技术和生态环保紧密结合。这已不再是未来的某个情景，今天我们就已经可以看到，建筑和设计正在向这个方向转型。

从我的角度来看，即使我百分之百同意当时理查德的理论，但我讨厌＂可持续性＂这个词，它是作为失败设计的一个借口，一个说法，或是一个理由，其实它并不是一个技术问题。问题的关键在于我们不断满足自身需求的同时，应该把环境的要求也一并考虑在内。人们应该逐渐意识到人类是不能无休止地向环境索取，不管其是否能承受。相反，大自然已经告诉我们什么是可以或不可以的，否则会造成不可想象的严重后果，最终导致地球上将不会有生命存在。如何让人们认识到在试图追求他们想要的生活的同时了解环境的必要性，我相信，未来这些问题将会成为社会发展的重大课题。

回到建筑，我经常与我的学生讨论他们的作品，他们的作品也曾发表在杂志上。学生们非常着迷于作品的形式和材料，对我来说建筑的外观并不是首位的，我的重点是如何在设计项目的时候学习到些什么。毫无疑问，在技术和材料的支持下，几乎任何事情都是可能的。问题的关键是如何融合设计、科技、社会、传统、文化、政治和甚至更多的元素。我相信，只有整合所有这些元素才能帮助我们创造一个成功的＂可持续性＂的项目，藉此应对21世纪的挑战。

建筑设计肯定不能改变世界，但它可以帮助人们意识到问题所在，并试图引领人们进入一个正确方向并最终找到解决方法。我相信年轻一代的设计师和建筑师们是可以做到这一点的。信息不断流通，如果在中国有任何有趣的项目，世界各地的建筑师们会在几秒中内看到并了解其中的内容。网络连接着全世界的人们并试图满足不同的城市需求。城市的数字信息规模已远远大于其传统结构。因此，信息流通已经非常的高速，这将促进文化交流，并在同一时间帮助人们意识到问题并提供解决方法。我认为这些技术作为一种工具，它为建立环保意识和文化交融提供了一个很好的机会。

我看了很多参赛作品，其中有好有差。我很惊讶只有部分作品能够贴近此次大赛的主题。今天只注重设计已经远远不够，没有与技术结合的设计将会过时，但是任何事情都需要时间，我相信再过15年，将不会有大赛再需要把绿色设计作为大赛主题，这将成为设计必不可少的元素。

最后我预祝此次活动能够圆满成功！

It has been a big pleasure to enter modern Chinese culture this way – through judging a ADPC design award. I am amazed about the general very high quality the projects have, however what is a big pity is the very apparent lack of sustainable thinking and doing. And this is a real pity since many projects easily with the right advisors could become sincere and very successful projects in terms of saving resources and using interior wise healthy materials and also making sure that spaces, products etc. They use as little energy as possible when needed.

One of the things I find very intriguing and also very interesting is the design based on old Chinese thinking and philosophy. The Chinese history is so rich and full of amazing examples of superior designs, use of materials, constructions etc. It is a great pleasure to see how many are actually applying, being inspired and making new interpretations of these sometimes ancient principles. I am in no doubt that Chinese designers and architects will bring new wonders to the world the coming decades – and it will be very exciting and stimulating to see a new and fresh influence on the global scene of design and architecture – similar to what has been seen happening in Japan and other nations in the far east. Design wise a country like Denmark has been seriously influenced by both Chinese and Japanese traditions – and I am personally very enchanted by oriental design and traditions – new as old. So therefore I would strongly suggest that Chinese designers and architects really profoundly start to implement sustainable principles in their thinking and doing. Having worked only with these themes for more than 25 years I can tell everyone that it's not only invigorating, fun, exiting but also the biggest and most interesting creative challenge the world community is confronted with – how do we design and develop a truly sustainable and comfortable world – for everyone on the planet. It's no small task, but it's evident to everyone now – and it's a must for making prosperous business – and developing the quality in life we all want. The world community needs to reduce massively its consumption of so many things – WE NEED TO THINK RADICALLY DIFFERENT – ITS NOT ENOUGH WITH A BIT OF SHIFTING MATERIALS. LIGHTS ETC. We all need to RADICALLY SUSTAINABLY INNOVATE – we need to come up with solutions for a comfortable life on this planet for more than 10 billion people – having a living standard that is significantly higher than what most of the world population has today. This is NO SMALL TASK, but as I have stated earlier for me it's the most exciting and interesting and it only becomes more and more challenging for every day seeing all the new materials, technologies, philosophies, services, and applications etc. being developed to support a global prosperous sustainable advance. We can make new and very exiting experiences in every possible category in society – globally. China is a new super power and it could become even more superior if it leads the way – especially in the new and RADICAL SUSTAINABLE INNOVATIONS its desperately needed and with China's incredible development and advancement in technology – It can also push things when it comes to design, architecture, interior design etc. And yes lots of projects are happening in China already – just we have missed them in this competition. So we from the jury are more or less in agreement about that this year's since the competition shows a profound lack of environmental awareness – we know it exists in China, you all just need to collaborate more. Therefore we hope we will see a lot of new and far more radical solutions for next year's awards – but not only from China but from the entire world.

Article

评委序言

Niels Peter Flint

很荣幸通过对APDC设计大赛参赛作品的评审，了解了现代中国文化。对参赛作品的高质量我感到非常惊讶，不过缺乏可持续设计的想法和应用使之成为遗憾。然而最大的遗憾是，很多设计具有很好的概念，如果能够节能，运用健康室内材料，同时确保空间、产品等的合理优化使用，将会成为非常成功的项目。

让我感到很有兴趣的是基于中国传统思想和哲学的设计。中国具有悠久的历史，有着无比卓越的设计、建筑和材料运用等方面的先例，令人欣慰的是，有许多参赛作品受其启发，古为今用，并使这些古老的哲学得到了新的诠释。毋庸置疑，中国的设计师和建筑师们会在将来的几十年中给世界带来新的奇迹，能够看到世界设计和建筑的舞台上注入的新的元素和力量令我感到兴奋和鼓舞，类似于日本和其他中东国家所发生的变化。这样像丹麦设计先进的国家已经受到了中国和日本传统文化的深远影响。我个人非常着迷东方的原创设计和传统——新就是旧。因此我强烈建议，中国的设计师和建筑师真正深刻地了解可持续理念并将其融入在你们的思维和行动中。

我已经致力于这一领域超过25年，我可以告诉每一个人，它不仅振奋人心，也是国际社会所面临的最大也最有趣的创意挑战。我们如何设计和发展真正的可持续和舒适的世界，对地球上的每个人来说是个不小的任务，但每个人都目睹其已成为商业关注的焦点，成为提升我们所向往的生活品质的必然趋势。国际社会需要大量减少对许多物质的需求，我们需要运用与以往截然不同的思考方式，仅仅转换材料、灯光等是远远不够的。我们都需要从根本上进行可持续创新，我们需要为在这个星球上超过10亿人口的舒适生活拿出解决方案，大多数人的生活水准将比现在高很多。这是个不小的任务，正如我之前所说，这是一个最令人兴奋和感兴趣的挑战。它会随着新的材料、技术、理论、服务和应用等方面的产生变得越来越具有挑战性。我们可以在社会每一个可能的领域创造崭新和令人兴奋的体验。

中国是一个非常强大的新生力量，随着令人难以置信的经济发展和科技进步的速度，对可持续发展创新的迫切需求也在上升，这也会推动涉及室内设计和建筑领域的方方面面。我相信有一些可持续设计项目正逐渐在中国产生，只是没有在这次大赛中体现，所以大赛评委或多或少会觉得今年的参赛作品缺乏环境意识。因此，我希望能在明年的大赛中看到更多来自中国和世界各国更新颖及具有创新精神的参赛作品。

Article
评委序言

Patrick Fong

Interior design is a reflection of society; from the early periods of Egypt, Greece, Rome, England, France, Italy and down the ages, to present-day China.

In the last twenty years, China interior designers and design have gone through some remarkable changes, as both respond to the evolution of our dynamic society.

Today, China's society is much more learned, demanding, and highly sensitive to issues of sustainability, environment and global exposure; and it is becoming more vocal.

I would like to congratulate the winners and sincerely thank all those who made a submission. Your outstanding work will not go unnoticed.

On behalf of the adjudicators and The Hong Kong Interior Design Association, I would like to convey my gratitude to the organisers , who have devoted long hours of hard work to make this award come truth.

I hope you will enjoy sharing with me the pleasure of browsing through the excellent works submitted in this year competition.

Professor Patrick Fong

室内设计是一个社会的映射，从早期的埃及、希腊、罗马、英格兰、法国、意大利，随着时代的变迁再到现在的中国。

在过去的20年中，中国设计师们和设计已经有了许多显著的变化，并随着社会的不断发展而不断进步着。

今天，中国的社会对可持续发展和环境问题高度敏感，有了更多的了解，也有更强烈的需求。

我在此向获奖者表示祝贺，并衷心感谢所有提交作品的设计师们。你们杰出的作品将不会被忽视。

我代表评委和香港室内设计协会，感谢本次大赛的组织者，他们为大赛倾注了长时间的辛勤工作。

我期待你能与我共同关注和分享本次大赛的优秀作品。

Creative Thinking

大赛特刊编委会
Editorial Committee of the Awards' Special Edition

编委会专家
Editorial Committee Experts

Claude Bérubé（加拿大）
国际室内建筑师/设计师团体联盟（IFI）历任轮值主席、亚太设计中心（APDC）主席。

Cheryl S. Durst, Hon（美国）
国际室内设计师协会（IIDA）副秘书长/首席执行官。

Jooyun Kim（韩国）
国际室内建筑师/设计师团体联盟（IFI）执委，韩国建筑师/室内设计师协会主席。

Michael Schwarz（德国）
迪拜阿扎曼大学建筑系教授，德国建筑师。

Ken Yeang（马来西亚）
马来西亚著名建筑师，国际建筑生态环境专家。

Patrick Fong（中国香港）
香港室内设计协会前会长。

Niels Peter Flint（丹麦）
国际著名设计师和艺术家，可持续发展的先驱。

Kevin Low（马来西亚）
著名建筑师，曾多次获得意大利、西班牙等国的设计大奖。

Philippe Moine（法国）
法国圣艾蒂安联合国教科文组织设计之都代表，著名设计师，曾获法国Conours Lepine 设计奖

Dietmar Muher（德国）
著名建筑师，德国柏林联合国教科文组织设计之都代表。

陈硕（中国）
英国建筑能耗标准评议员，2010上海世博会零碳馆馆长，零碳中心总裁。

陆耀祥（中国）
上海交通大学建筑工程与力学学院、建筑设计研究院院长。

编委会顾问
Editorial Committee Consultants

徐凌志（中国澳门）

澳门中西文化创意产业促进会会长。

何增强（中国）
上海创意产业中心秘书长、上海国际创意产业活动周秘书长。

李水林（中国）
中国杭州文化创意产业博览会组委会办公室主任。

黄浩然（中国）
无锡市文化艺术管理中心副主任、无锡博物院院长。

黄永（中国）
法国尚飞中国分公司总经理

编委会成员
Editorial Committee Officers

总策划：王 怡
协 调：王 斌、徐 斌
策 划：金正宇、魏振飞
编 辑：于 勤、庄震华
平 面：刘 婷、钱佳妮
翻 译：徐 佳、余佳瑛

目录/CONTENTS

项目名称：Teman Sejati Sarihusada 中心
所 有 者：PT. Sari Husada
地　　点：印尼日惹
室内设计：Ayu Sawitri Joddy

作为HCP强力推荐的品牌制造商PT Sari Husada，建造了TSS中心，把它作为公司社会责任的一部分。这样做是为了回报其重视的顾客（SGM和Lactamil），项目提供了儿童活动室、哺乳室、娱乐区、营养咨询室、办公室和前台区域的公司介绍摊位，还有哺乳区/尿布更换区。这个地方对于儿童、妈妈和准妈妈而言都是个绝好的地方。因此，一个精通儿童和母亲营养学的让人信任的专家是必要的。所以，让TSS中心与其社会更接近、更紧密，同时提高所需的口碑来发展项目。

此项目是2010年HDII获奖项目中最终挑选的26个之一，已在ASRI杂志上发表过。

藤条吊灯已从2010年委员会的desain.id中获得了gooddesain.id标签。

这有一些可持续性的方法应用于此项目：

caterpillar休闲长椅和藤条做的cocoon吊灯；

我们用水葫芦箱子而不是塑料盒为孩子们作玩具箱；

我们的家乡工作室位于jaharta，但自从此项目定居jogjakarta后，我们使用的工人和材料均来自于jogjakarta市周边地区，而且我们的内置家具大多数在jogjakarta生产；

我们利用当地资源，使用kamer木作家具表面，而非柚木；

我们把此遗产建筑的大翻修的程度减到最小，保留了老式瓦片、门、窗框和木板等；

我们在每一间教室里安装了新窗户以保证白天有充足的阳光；

我们在大厅使用吊扇来代替空调；

我们改变了下客区的功能，将其变为一个开放的区域用做等候区。

PT Sari Husada as producer of brands most recommended by HCP, built Teman Sejati Sarihusada Center as part of their Corporate Social Responsibility, They build this Center to spoil their loyal consumer (SGM and Lactamil), it provides kids activity class room, lactation classroom, play ground area, nutrition consultation room, office, and representative info booth at reception area and also breastfeeding area/diaper changing area. It's a perfect place for children, moms and moms to be. So it must deliver assurance of a trusted expert in kids and mother nutrition. There for bringing the TSS Center closer and more intimate with its community and at the same time strengthens word of mouth needed to grow the program.

here are some of the sustainable approach taken on the project:

- the caterpillar waiting bench and the cocoon hanging lamp made of rattan

- we use water-hyacinth box instead of plastic box for children toys boxes

- our home based studio is located in jakarta but since this project location in jogjakarta, some of the labor and materials came from around jogjakarta city and we produced the built-in furniture mostly in jogjakarta

- we're using kamper wood for facade from local source, instead of teak wood

- we minimize the major renovation of this heritage building, we keep the old tile, old door/window frame and panel, etc

- we made new window in one class to maintain sunlight as a daylight

- we use ceiling mounted fan in the lobby instead of AC

- we change the drop off area function as an open-plan area for waiting area

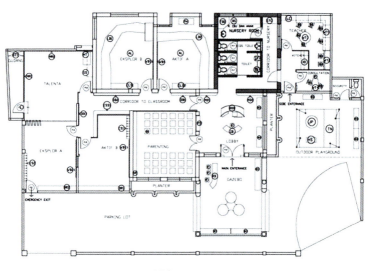

平面布置图

商业空间
Commerce

金奖 / Gold Award

②

室内设计理念

作为新加坡知名的金融集团，大华银行在2010年为旗下"Privilege Banking尊享理财"颁布了全新的企业识别系统并全面开展在中国的VIP金融服务。为了铭记这个历史性的一刻，大华银行决心打造一个地标性的建筑物，它将成为新天地乃至整个上海滩的聚焦中心、富豪和精英们的理财顾问首选。为了帮助业主将这个理念付诸实现，设计师打破了传统意义上银行设计的窠臼，呈献给业主一个融合银行及理财服务与精品酒店氛围的全新设计概念。希望除了银行的功能以外，这个空间能成为上海商业领袖和文化精英们聚集的财富中心。

在传统的中国文化当中，风水与财富有着密不可分的关系。"水"，象征着财富的"来源"，而"珍珠"则代表了财富的"结果"。无论对于大华银行集团及他们尊贵的客户来说，"水"和"珍珠"，赋予了绝妙的寓意。因而，不管在外建筑立面还是室内空间设计中，我们都透过含蓄而优雅的表达方式，来诠释着"水"与"珍珠"的概念。

首先是银行的外立面改造设计，我们摈弃了原建筑传统的大理石装饰，全新设计了特殊效果的玻璃幕墙。主体是由定制钢化蓝玻璃经过喷砂雕刻所制成。视觉效果的水纹图案，则由喷绘艺术家直接在玻璃幕墙上描绘抽象的"水波荡漾"画卷，每片玻璃图案均有所不同。多层次的蓝色帷幕墙透过全幅镜面不锈钢背衬，在白天呈现出犹如珍珠贝母一般的色泽。而在夜晚，当暮色降临，在LED灯光的变换下，整个建筑物焕发出如同深海的湛蓝，令人不自觉地驻足欣赏。这个夺人眼球的独特建筑处理，使得大华银行"尊享理财"中心在奢侈品旗舰店林立的新天地，占有一席之地。

进入到室内大堂，从天花造型到主体墙，富有曲线和韵律的造型设计，和室外建筑设计融为一体。大厅等候区的上方悬吊了精致的水晶吊灯，宏伟的背景墙上，巨幅的荧幕及多媒体讯息屏提供即时的国际财经资讯及全球股市行情，让"尊享理财"的客户在等待之余，随时掌握理财讯息。

穿过大堂区，步入贵宾区域，我们设计了别致的水景，水从天花错落的管中，优雅地落了在只有一厘米厚度的无边界水池中，配上象征水中珍珠的雕塑，就好像珍珠落玉盘，展现了一幅灵动的画卷。

乘上电梯进入二楼的私人贵宾区，不经意中眼前一亮，我们为更高贵的"尊享理财"贵宾客户设计了一个非常高雅而舒适的尊享酒廊，可以想象一下，来本银行惠顾的精英人士，端着香醇的咖啡，舒适地坐在沙发上，迎着阳光的沐浴，享受着最为高端的咨询服务。到了夜晚，这个空间在点亮光纤帘及情境壁炉的烘托下，立即能转换为贵宾精英会所，顶级的珠宝鉴赏会、高级理财讲座及顶级评酒派对都是为高端客户量身定做的社交礼遇。

眼前的一切，不仅帮助业主实现了他们最初的梦想，也创造了一个设计业界的新探索，开创了高端金融服务的全新视野。

United Overseas Bank China also fondly known as UOB has broken out of the typical mould and boldly embraced the combination of privilege banking and the luxury of a high-end boutique hotel

项目名称：新加坡大华银行上海新天地支行及"尊享理财"中心
项目地址：马当路156-158号
主设计师：Sunny Wang、Liza Shi

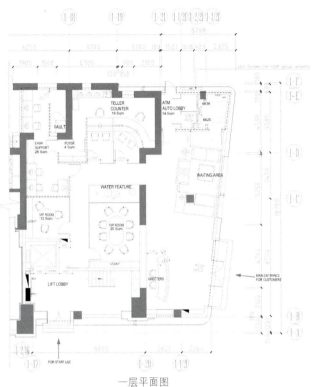

一层平面图

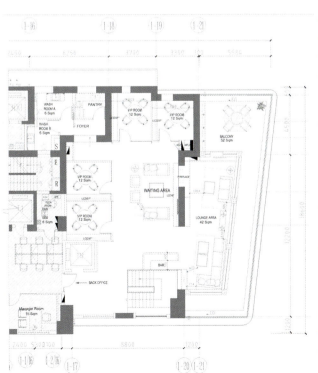

二层接待区平面图

concept proposed by their Interior Design Consultant, SLA. It was designed to accommodate not only the bank's functions but also have the lavish facility to hold events for their elite clients.

SLA, conceptualized the interior space and exterior façade using water element and the precious pearl where in Chinese tradition, it is associated to a source of wealth. These elements were subtly designed into the exterior glass façade and interior space.

Being in the exclusive location of Xintiandi, the façade was given a new face. It went through a custom sandblasted and artist abstract painted blue glass wall giving it a vision of waves. In daylight, with the mirror finish stainless steel background, it reflects the colours of the mother of pearl and when dusk falls, the system controlled LED lights illuminates the colours of deep blue sea.

Moving into the interior space, the lobby carries through the wave abstract design on the ceiling and backdrop with different use of materials. This space combines the luxury of crystal chandeliers and the technology of wall-to-wall digital multimedia.

Walking through the corridor into the VIP meeting rooms, you will be greeted by a sleek and borderless water feature with pearl sculptures representing the message of water flowing onto a jade platter accumulating fortune.

The luxurious VIP space is then extended to level 2, where the elite clients receive high-level consultation service. A spacious lounge area was intentionally designed to allow the bank to hold private functions for their clients. The windows panels here are treated with fiber optic curtains where it provides the privacy within and maintains the view to the outside.

And it is here, UOB Privilege Banking Center makes her iconic mark in Shanghai!

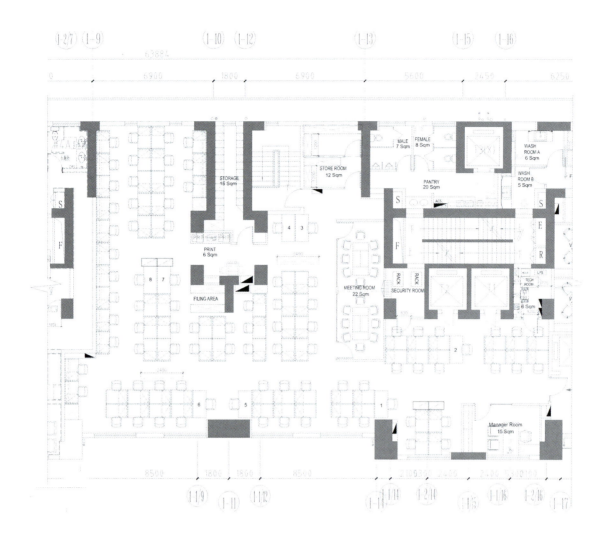

二层办公区平面图图

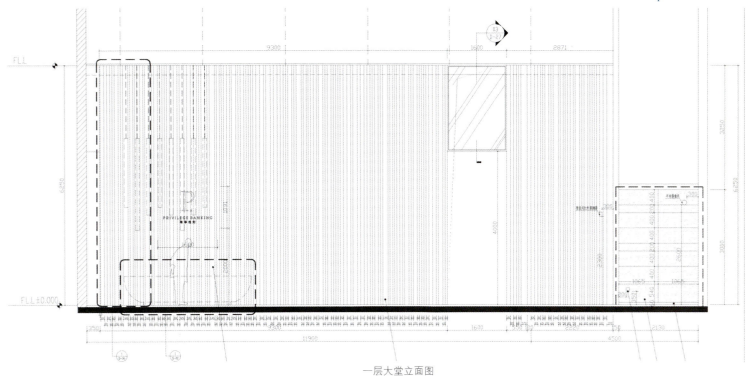

一层大堂立面图

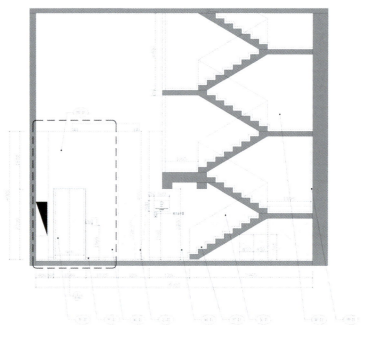

一层电梯厅立面图

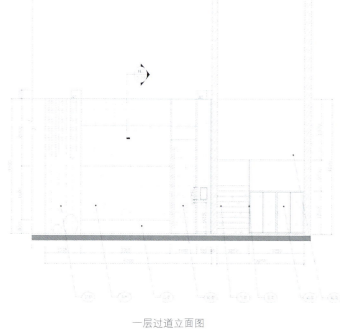

一层过道立面图

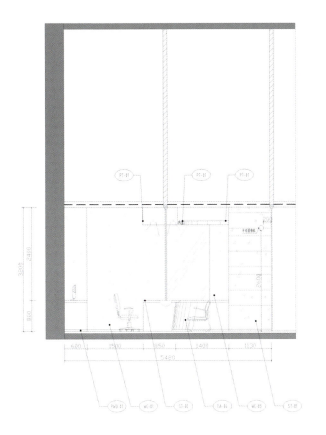

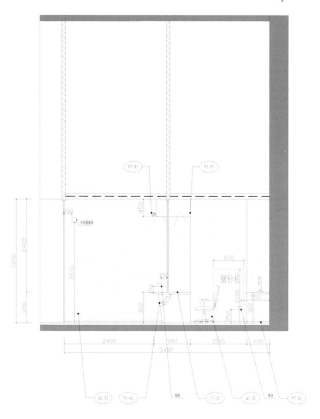

一层现金柜立面图

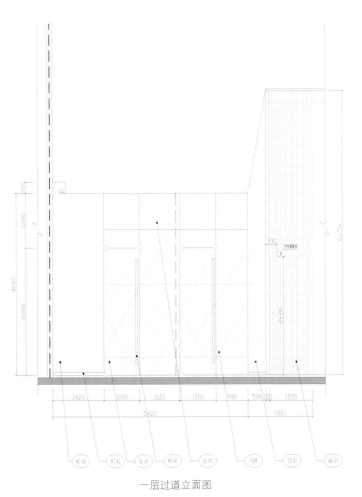

一层过道立面图

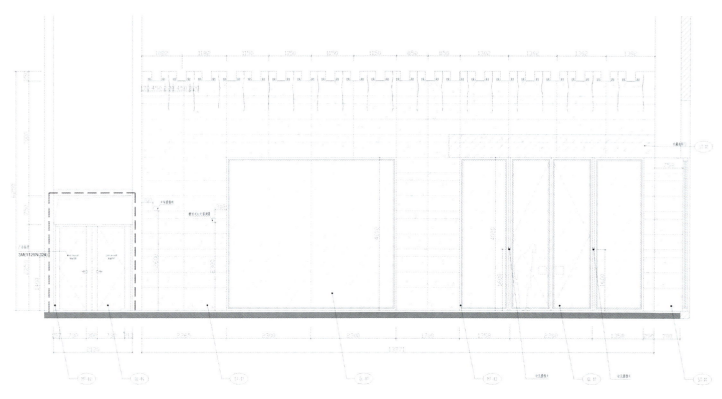

一层大堂立面图

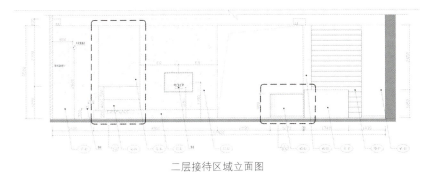

二层接待区域立面图

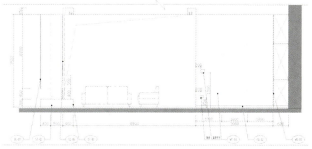

二层接待区域立面图

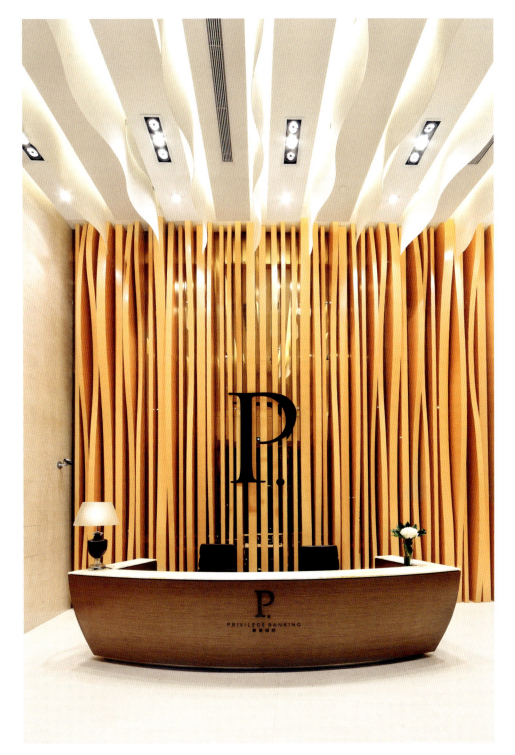

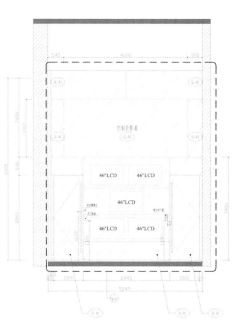

一层大堂立面图

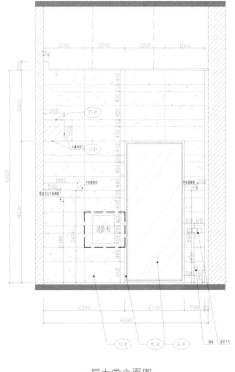

一层大堂立面图

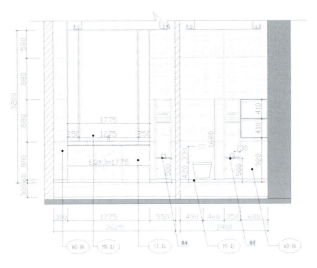

二层接待区卫生间立面图

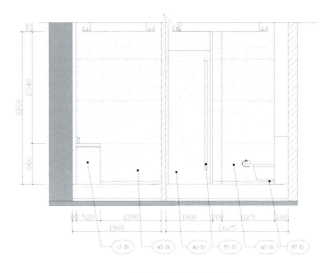

二层接待区卫生间立面图

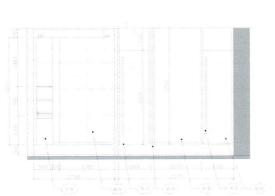

二层接待区茶水间立面图

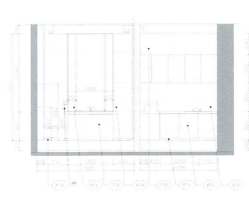

二层接待区茶水间立面图

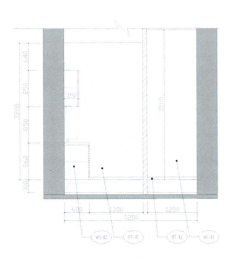

二层接待区茶水间立面图

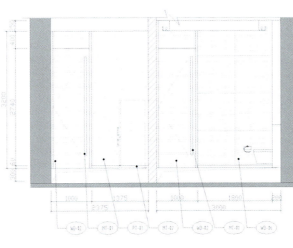

二层接待区茶水间立面图

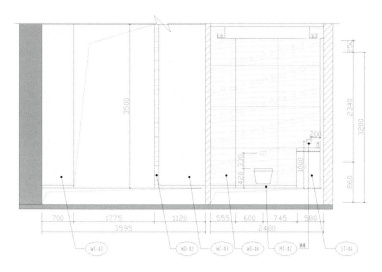

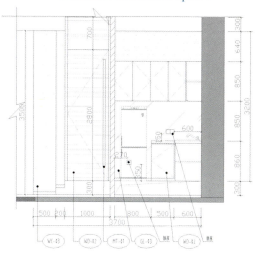

二层接待区卫生间立面图   二层接待区卫生间立面图

商业空间
Commerce

 银奖 / *Silver Award*

3

项目名称：一清茶事
设 计 者：陈谷
设计单位：红星空间生产队

东情西韵一清茶事

知堂老人说：喝茶当于瓦屋纸窗下，清泉绿茶，用素雅的陶瓷茶具，同二三人共饮。得半日之闲，可抵十年尘梦。一清茶事恰是为想偷半日之闲之人所等候的地方。

设计师陈谷一直在想，做个真正低碳的物业，可以是小小个性的，可以是高调大众的，关键是可以完全可以用原始自然和低碳的材质来完成设计。做自己想做的，玩自己想玩的。有缘于一朋友，才得有今天的一清茶事。

在家乡安吉，"一清茶事"以自己完全独特清雅的姿态，静观周边城市的喧闹。门庭前是莲与竹相伴，棕榈鸟巢屋顶、原木大立柱、鹅卵石砌门墩。远观所见，皆是原生态自然之物，亲密入眼。

做了快10年的设计装修，越来越感觉自然、生态材质之美，犹如看赏一浅俗美女，外表惹眼的自是吸引人，但不能开口说话、交流，一开口便是泄了气的皮球，因无内在的气韵，美，也只能是外在的、一时的。而如旧石器、泥、砂、鹅卵石、竹、原木等一切的自然生态之材越来越经得起时间的推敲，不会随时间的推移而有厌恶之感。一清茶事的装修中，摒弃了最常用的装修材料，一切尽可能地利用可再生的自然之物，断断续续收集来的老青砖，被别人称之为破木头的老旧木板、旧石器，这次终于有机会大放光彩。老木、旧石的大量直白的堆砌会有深沉的腐朽之气，反倒会抹杀了它们的美。老青砖走廊隔断墙洞里的风情烛台相应衬，飘逸的纱缦来配硬冷的石器，说是自己的原创也罢，只是在这样玩的过程中组合成了最美的视觉效果。茶楼的过道墙面是小时候老家的稻草泥墙，增加了些质感。地面用的是黄金砂（和小时候的石子墙有异曲同工之处）。

从原木门头延续而至的老青砖走廊隔断，稻草泥墙、手工竹丝灯笼、隔山隔水淘来的老旧石槽、石盆、石柱、石墩与轻柔的纱缦，梦幻的烛台相应衬，增添了无限的温情。茶楼异域风情门头和金色如意拉手相配，入口大厅的传统笔划吊顶、老青砖镂空过道墙、老砖水池、石钱水景、稻草泥墙、手绘宫灯纷纷显现出独特的魅力和原创性以及彰显了中国的传统文化。装修格局为二楼是传统古典中式包厢，三楼卡座是浪漫东南亚格调，杉木段不规则的排列就是一堵隔断，若隐若现的纱缦间，迷离的灯光下，不经意抬头，会一不小心撞有惊鸿一瞥。此间此景极适情人、朋友小坐。最奢侈的是凛凛冬日里温情的太阳，因门前开阔，阳光独好，可以捧一本好看的书，点一杯酽酽的茶，伴着好听的音乐，一个冬日下午就是完美的了。三楼最里间的是隐密的禅茶室，赤足盘腿而坐，可以独自一个在此袒然静思，静化心灵，也可与老师、朋友交流品茶与养生之道，感悟人生。

Something of tea—domestic and exotic experiences

Mr. Zhou Zuoren said, it is best to drink with couple of friends near historic house with tile roof and paper-made windows. The green tea leaves need be brewed in spring water and carried in simple ceramic tea set. The rustle and bustle in ten years will be removed according to the suspension of worrying in this half day. "Something of tea" is just such a place for those who want to have half a day off.

Chen Gu, a designer, has been thinking about starting the business on the basis of Environmental Protection, no matter how distinctive it should be, no matter how popular it would be, The important thing

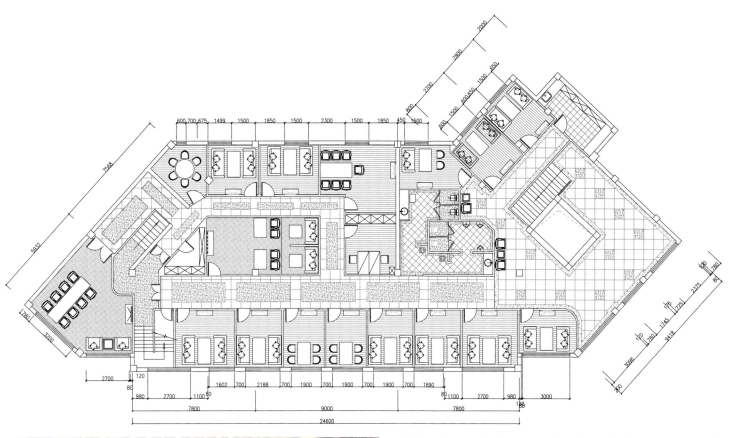

is that design can be totally completed by natural and environmental-friendly materials. Doing business which he wants, enjoying life what he dreams, and making friends who he likes. Finally the "Something of Tea" is coming into being.

"something of tea", which is located in Anji, one is able to observe the clamor while maintaining a cool and elegant posture. Before the door, the lotus and bamboos grow together and palm nest can be seen in the roof. There also exists a log crutch and cobblestone-made rock island. From a distance, all of them look so intimate to the eye, primitive and natural.

After ten years' experience in interior design and decoration, he becomes better able to appreciate the beauty of materials that from nature. Take a beauty as an example, although her appearance is alluring, her words and thoughts are too poor to show her inherent quality. Therefore, the external beauty is only superficial and temporary. With compared, all the natural and ecological materials such as old stone implements, mud, sands, cobblestone, bamboo, log and others tend to stand the long time and won't be abhorred as the time goes by. We try our best to utilize the reproducible natural materials instead of the common ones when decorating "something of tea". Old black bricks collected over time and the old wooden boards and stone implements which people called garbage finally have their chance to stun the world. If an awful lot of old wood and stone are piled up directly, they will looks dull, and, their beauty may be negated. The wall opening of the partition in the old black brick-made aisle should be with romantic candlestick while the hard cold stone implement should be matched with flowing veil. It can be said of my own creation. The wall in the aisle of teahouse is the mud wall commonly seen in hometown in childhood, which enhances the sense of quality. In addition, the floor is made of golden sand, which is suite similar to the pebble wall in childhood brought to us.

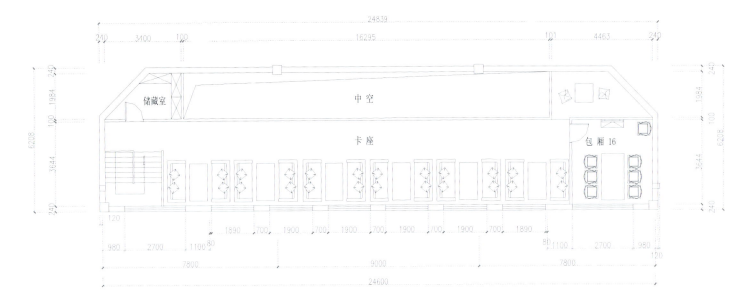

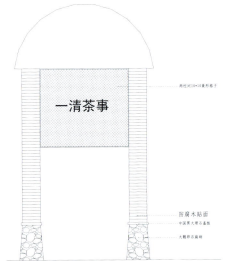

the partition of the old black brick-made aisle extending from the log gate, the mud wall, the bamboo sticks hand-made lanterns and the collected old stone cistern, stone basin, stone pillar and stone seat all veiled in light gaze and dreamy candlelight, the place is filled with tender and sweet ambience. Unique charm, originality and Chinese culture of this tea house are distinctively shown in its exotic amorous door with golden ruyi handle, the traditional handwritten pendent lamp in the entrance of hall, old hollowed-out black brick corridor wall, old brick basin and hand-painted palace lanterns (it may not excellent because of lacking practice) When it comes to the layout of decoration, the rooms on the second floor are traditional classical Chinese ones; by contrast, the third floor's are romantic southeastern style, where long and short standing China fir compose a

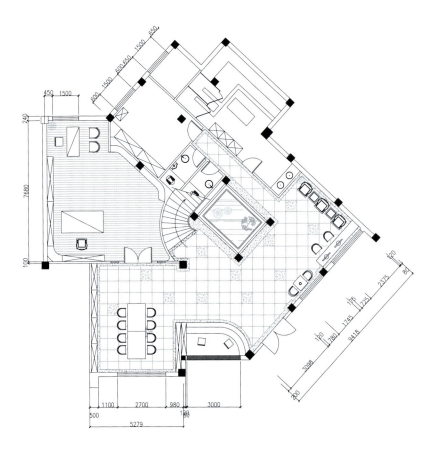

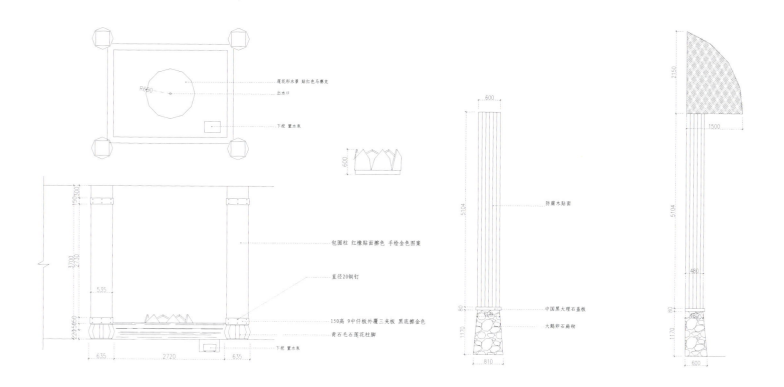

蓮花形水景 贴红色马赛克

出水口

下挖 置水泵

防腐木贴面

包圆柱 红橡贴面擦色 手绘金色图案

直径20铜钉

中国黑大理石盖板

150高 9中仟板外覆三夹板 黑底擦金色

青石毛石莲花柱脚

大鹅卵石扁砌

下挖 置水泵

50x20厚平板线条擦色
墙面石碴拉毛刷色
25x25线条刷白
中通透

木花格 后衬羊皮纸
4厘米宽线条压边

面板擦色
玻璃层板

120高木质踢脚内进2厘米

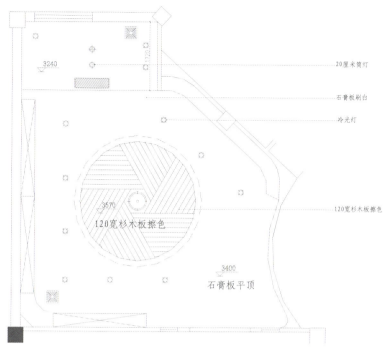

20厘米筒灯

石膏板刷白

冷光灯

120宽杉木板擦色

120宽杉木板擦色

3570

3240

3400

石膏板平顶

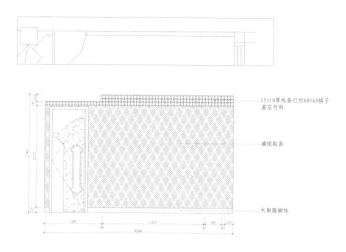

15×10厚线条打约60×60格子
基层竹帘

墙纸贴面

木制器脚线

partition. Under the blurred light, some may raise heads, often by accident, meet with others from the looming veil. This kind of room is most suitable for couples and friends. The nicest thing is the warm sun in winter. Because of the spacious porch, one can enjoy ample sunshine here with a cup of tea and pleasant music, it would make a perfect winter afternoon. The innermost clamber is a hidden Chan tearoom, where people can sit across-legged barefooted tranquil solitude to purge the mind. Perhaps, it would be also a good choice to have a chat about a way of drinking tea and keeping healthy as well as something of life with teachers and friends.

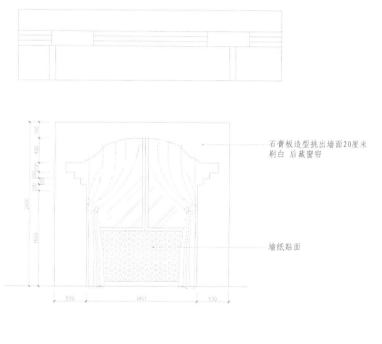

石膏板造型挑出墙面20厘米
刷白 后藏窗帘

墙纸贴面

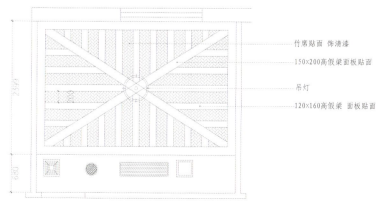

竹席贴面 饰清漆
150×200高假梁面板贴面
吊灯
120×160高假梁 面板贴面

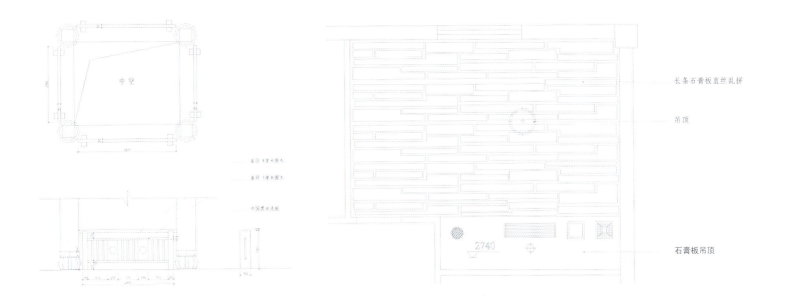

長條石膏板直絲亂拼

吊頂

中空

直径 8寸水圆木

直径 7寸水圆木

中国黑水点板

石膏板吊顶

2740

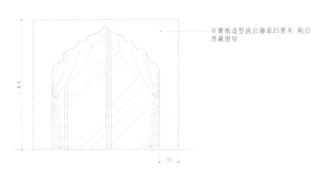

石膏板造型挑出墙面25厘米 刷白
厚藏窗帘

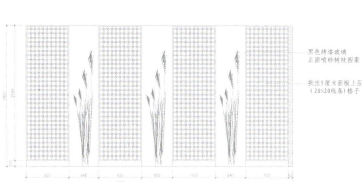

黑色烤漆玻璃
正面喷砂树枝图案

挑出5厘米面板上压
（20×20线条）格子

项目名称：阿哆诺斯蛋糕欧洲城店
设 计 者：葛佳亮

阿哆诺斯

本案是位于浙江省温州市高档住宅区和购物区非常显眼位置的一家欧式蛋糕店——阿哆诺斯。

一入店堂，高而宽敞的格局就让人心情舒畅，双圆形的吊顶设计，简约又富有层次感，颇为大气。米黄色的基础色调配以咖啡色系的货架，浓郁的烤培气息飘香而来，橱窗设计采用灯光变化，错落有致的摆放，勾起人的购买欲望。

在大厅蛋糕展示柜旁是天然的毛石墙面，红砖砌的壁炉的搭配，自然联想起欧洲电影的画面——欧洲乡村的老主妇在厨房忙着烘培面包招待远方的客人，烘托出整个卖场的气氛！

阿哆诺斯店旁的露天小桌，户外花栏配以休闲餐椅，阳光晴好天气坐在那里，可以欣赏一路的风景。

This case is a European-style cake – Adono's, which is located at a very prominent position in Wenzhou's upscale residential and shopping district.

As step into the shop, the high and spacious pattern makes people feel comfortable; besides, double-circular ceiling design, simple but very layered, showing its great generosity.Together with the cream colored basic tone and brown shelves, whole room is full of rich roasted flavor. Windows design used lighting changes, the patchwork of display help evoke customers' shopping desire.

Next to cake display cabinets standing the natural rubble walls, red brick fireplace bring people into the European movie pictures —— in which an old housewife is happily baking for guests afar, exactly consistent to the atmosphere people are involved in!

In addtion, open-air seats near Adono's, sitting leisurely on chairs among flower columns coupled with the sun in fine weather, you can enjoy the scenery and delicious food at the same time.

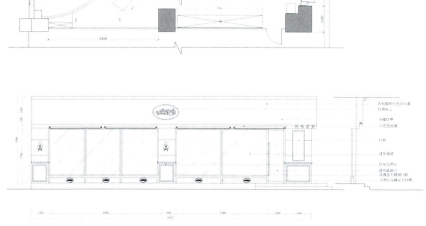

阿哆诺斯欧洲城A外立面图

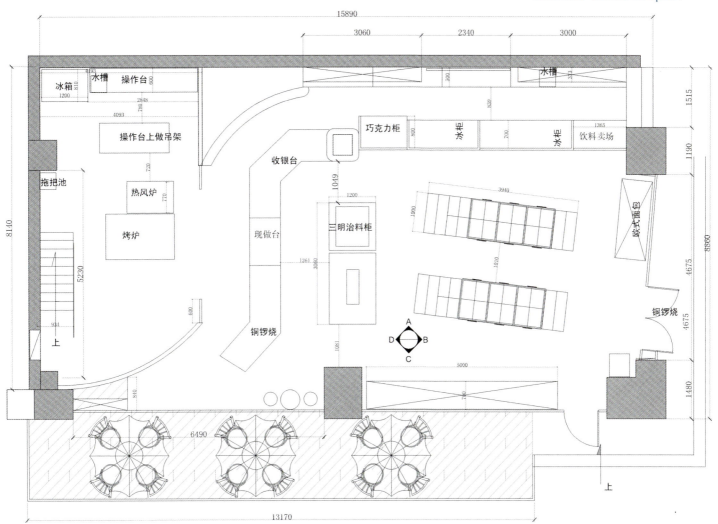

冰箱
水槽
操作台
操作台上做吊架
拖把池
热风炉
烤炉
上
收银台
现做台
巧克力柜
冰柜
冰柜
饮料卖场
水槽
三明治料柜
铜锣烧
欧式面包
铜锣烧
上

阿哆诺斯欧洲城平面图

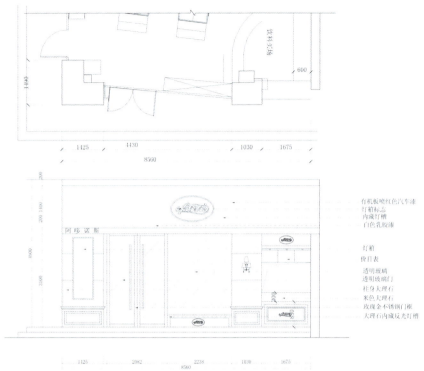

饮料火场

8560

1425   4430   1030   1675

有机板喷红色汽车漆
灯箱标志
内藏灯槽
白色乳胶漆

灯箱
价目表
透明玻璃
透明玻璃门
柱身大理石
米色大理石
玫瑰金不锈钢门框
大理石内藏反光灯槽

阿哆诺斯

8560

1125   2082   2238   1030   1675

阿哆诺斯欧洲城D外立面图

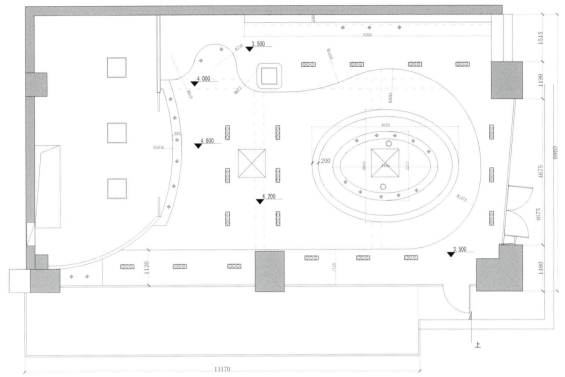

阿哆诺斯欧洲城天花板图

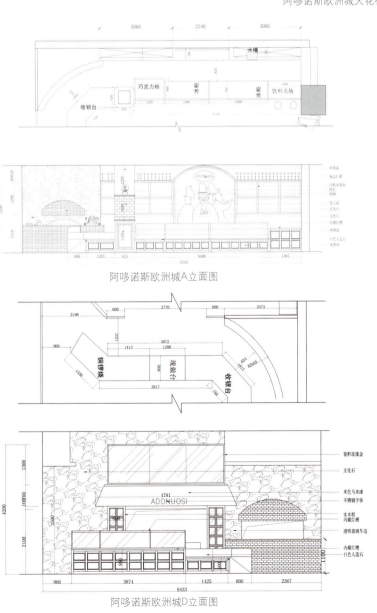

阿哆诺斯欧洲城A立面图

阿哆诺斯欧洲城D立面图

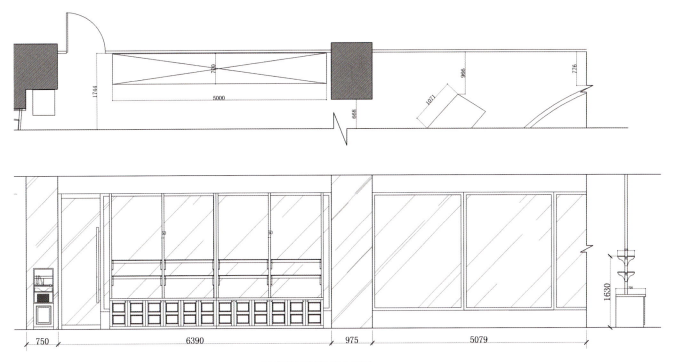

阿哆诺斯欧洲城C立面图

项目名称：杭州西溪MOHO售楼处
设 计 者：朱晓鸣
设计单位：杭州意内雅建筑装饰设计有限公司

自由的灵动

本案是展售中心的楼盘，针对的是80后上下从事创意产业为主的时尚群体。

结合本案项目所在地空间较为局促等几方面综合考虑，没有刻意地追求如何在小场景中创造大印象，如何跳脱房产销售行业同质化令人紧张的交易现场，如何创造一种氛围更容易催化年轻群体的购房欲望；而是将本案定位在纯粹、略带童真，甚至添加了几许现代艺术咖啡馆的气氛，作为切入点。

整体的空间采用了极简的双弧线设计，有效地分割了展示区以及内部办公区，模糊化了沙盘区、接洽区和多媒体展示区，使其融合在一起。

空间色彩大面积采用纯净的白色，适当点缀LOGO红色，吻合该项目的视觉形象。

智能感应投影幕的取巧设置、树灯的陈列，还有开放的自由的水吧阅读区，综合传递给每位来访者这样一种感受：轻松而又自由、愉快而又舒畅、畅想而又提前体验优雅小资的未来生活，这也完美表达了MOHO品牌的内在含义：比你想象得更多。

Free Motion

This design case, real estate sales centre, targeting generation X's working within the creative industries. Combining this case in a relatively constricted design space we had to think of many solutions, how to create a big impression in a small space, how to break through the classic real estate office design and make people feel less anxious about purchasing. How to make an atmosphere that will induce this younger generation to desire to be a home owner. The answer we came up with was to turn to purity and childhood simplicity and at the same time adding many modern additions such as art café style as a breakthrough point. The whole space uses very simple double curve designs, which is effective in segregating the display area and inner offices and at the same time obfuscating the building model, media and reception areas.

The color scheme of the space uses large areas of pure white which provides excellent contrast to use a red LOGO; it compliments this projects visual imagery.

Intelligent inductive projected movies, lighting trees, open plan water bar and waiting room provided for the young waiting customers. Young, carefree, free thinking and happy about their new fantastic lives ahead of them, this perfectly embodies the MOHO brand slogan "Even better than you can imagine".

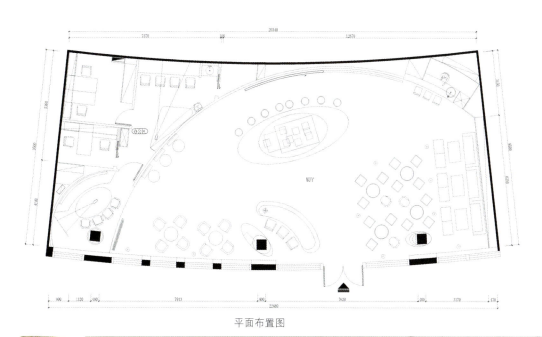

平面布置图

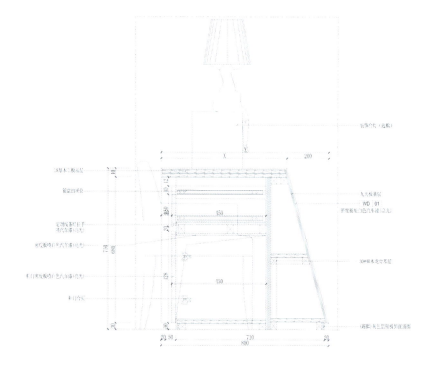

接待台平面图   节点大样图

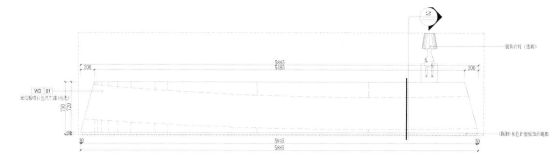

接待台正立面图

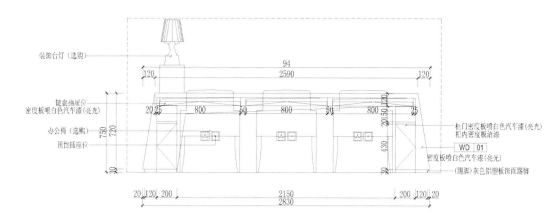

装饰台灯（选购）

键盘抽屉位
密度板喷白色汽车漆(亮光)

办公椅（选购）

预留插座位

WD 01

柜门密度板喷白色汽车漆(亮光)
柜内密度板清漆

密度板喷白色汽车漆(亮光)

(踢脚)灰色铝塑板饰面踢脚

接待台背立面图

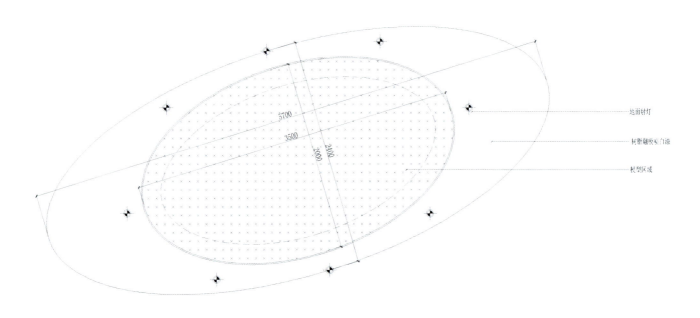

地面射灯

树脂磨块刷白涂

模型区域

沙盘平面图

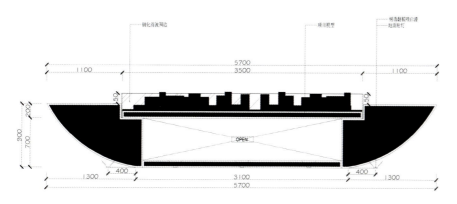

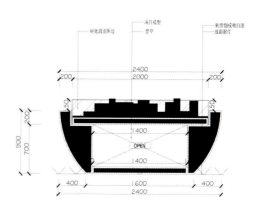

沙盘平面图

德国ROY洁具（南京旗舰店）

本案的设计理念来源于产品本身，洁具本身的洁净、透彻，犹如冰川般"宛若天成"。设计中将"海洋的浮冰"引入整个空间当中来，产品被从海面的"浮冰"中"飘"出来。波浪起伏的外墙、热弯的玻璃、忽高忽低的"冰川"墙体、"漂浮"不定的展台，产生很好的趣味性，并很好地诠释了产品自身的特性。

German Roy Vanities (flagship store)

The design concept comes from the products themselves. Vanities appear a clean, clear profile, like a nature born glacier. The "floating sea ice" is included into the space; products are softly pushed out from the "floating ice" in the "sea". The multi-level wavy external wall of sweat-out glass, the crooked "glacier" wall, and the "floating" exhibition stand are united together to show off a good taste, and naturally express the feature of the product.

项目名称：德国ROY洁具旗舰店
设 计 者：贾怀南、Enrico（意大利）
设计单位：上海半千舍建筑装饰设计有限公司

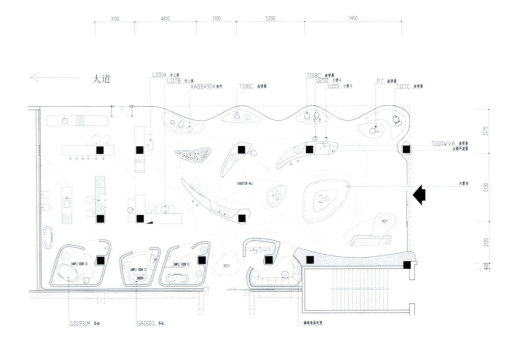

洁具展厅布置图

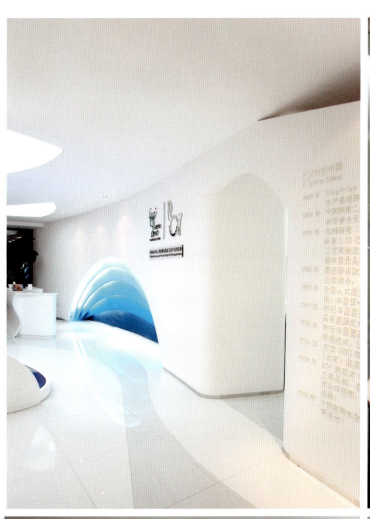

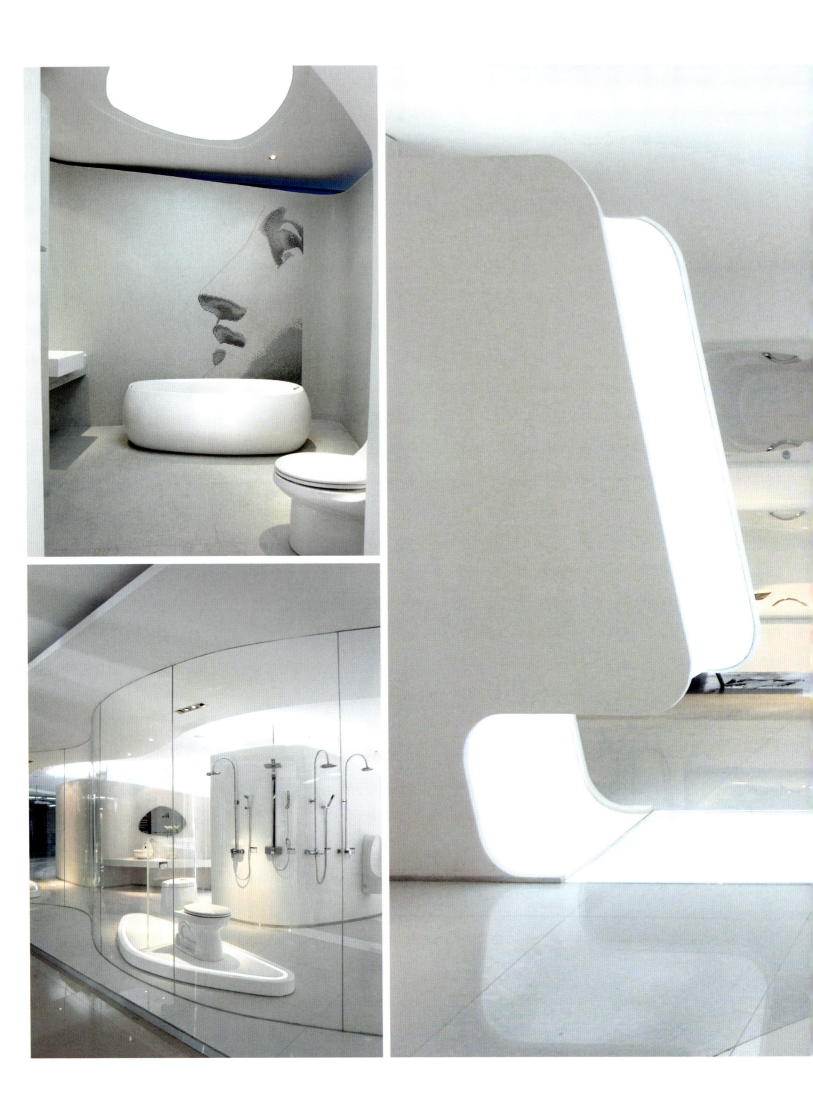

优胜奖 / **Winning Prize**

4

本案为西班牙进口瓷砖展馆，风格定位为新古典主义。

新古典主义兴起于18世纪的罗马，其宗旨是对巴洛克和洛可可的反动，其次以重振古希腊古罗马的艺术为信念……其排斥对以前文化无限制地复制和无条件地接纳，提倡采用新的工艺、新的材料、新的设计理念去表现内心对美的感悟，在注重装饰效果的同时，用现代的手法和工艺还原古典气质，新古典主义具备了古典与现代的双重审美效果，其完美的结合也让人们在享受物质文明的同时得到精神上的慰藉。

其实，我不想以某种风格去定位本案，期间参阅了季裕棠先生的设计风格并且确认了方向，我和甲方喜好平和、不喜欢起伏，希望可以在这个设计中传递着以静带动理念的空间，就像季先生所说：所谓奢华只不过是一种营销工具，就内涵而言，奢华没有任何意义

It is used as an exhibition hall for ceramic tiles(imported from Spain).The whole style bases Neoclassicism.

Neoclassicism rised in 18th century in Rome.The aim was against for Baroque's and Rococo's. People were faith at revitalizing the art of ancient Greek and ancient Rome.No cultural duplication,no unconditional acceptance for the previous culture.People were strongly encouraged to absorb new technology,using new materials or a new design concept to show up their bottom inner world.People

工程地址：杭州市西湖区新时代E座21号
户型结构：平层
使用面积：420 平方米
项目名称：欧汇臻品展厅
设 计 者：梁苏杭
设计单位：中国建筑学会室内设计分会杭州分会

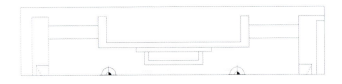

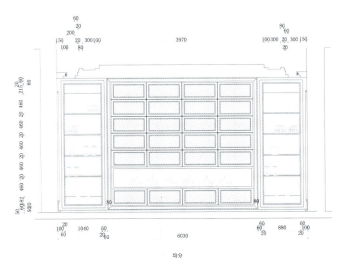

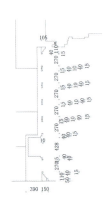

均分

立面图

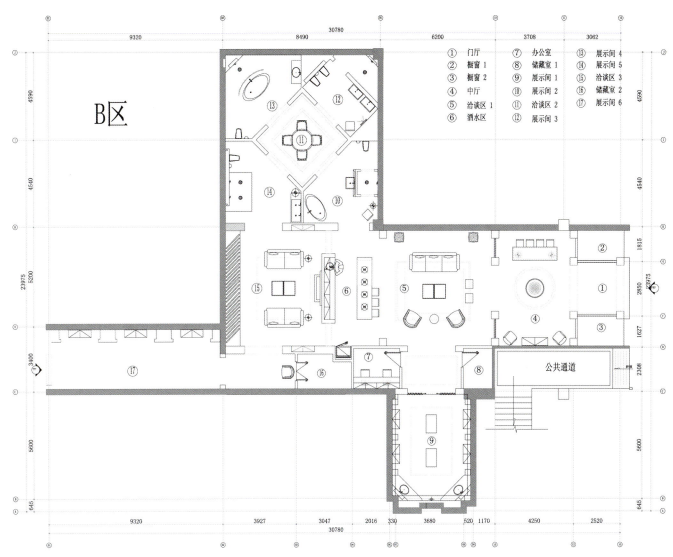

B区

① 门厅	⑦ 办公室	⑬ 展示间 4
② 橱窗 1	⑧ 储藏室 1	⑭ 展示间 5
③ 橱窗 2	⑨ 展示间 1	⑮ 洽谈区 3
④ 中厅	⑩ 展示间 2	⑯ 储藏室 2
⑤ 洽谈区 1	⑪ 洽谈区 2	⑰ 展示间 6
⑥ 酒水区	⑫ 展示间 3	

公共通道

平面布置图

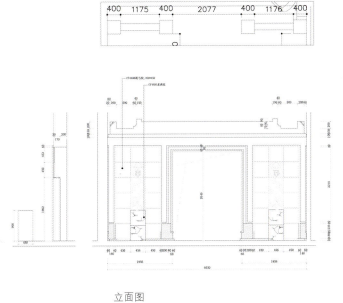

立面图

not only emphasized at decoration,but also restored the classics in a modern way. Neoclassicism conveyed the essence of both classicism and modernism,and the combination gave people substance satisfactory and spiritual comfort.

To be frank,I do not prefer to name the design as a given style.Before confirming the

direction of my design,I refer to some works of Mr Ji(Ji Yutang,a great master). My client and I all like tranquil,calm life,so the main idea of my design:keep a vivacious meaning in a quiet design,and I hope my design concept can be understood and can get approval.

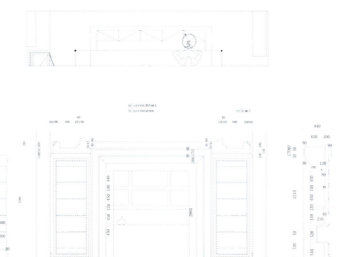

立面图

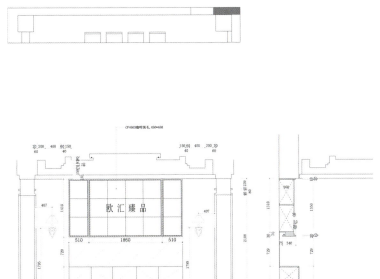

CF4503咖啡绒毛, 450*450

欧汇臻品

立面图

商业空间
Commerce

优胜奖 / Winning Prize

4

项目名称：奥维尔销售中心
设 计 者：熊錾
设计单位：成都主道空间设计工程有限公司

法式田园小镇

本案定位是高档售楼中心。设计师选择了梵高、塞尚等艺术名家曾长期停留过的地方——位于巴黎西北约30千米的瓦兹河畔的奥维尔小镇作为创作的原点。

奥维尔销售中心分为一、二两层。大厅是整个空间给人的第一印像，设计师结合了手绘油画和浮雕的表现元素，并与特别设计的定制吊灯相交融，营造出一种低调华丽的艺术生活品味。 一层奥维尔体验区打造的是一种半室外的感觉，融入了植物和法式田园小镇等元素。材质上以质朴的木地板以及墙面硅藻泥为主，并通过与室外光线的结合来营造氛围。 二层的书吧是私密空间的一个序列开端，它的光线不能太亮。对称的弧形墙共同指向酒水吧台后的植物墙。视线穿过法式小吊灯，扫过弧形墙，停顿在植物墙上，同时脑海里回想起刚才看过的书，嘴里品味着法国红酒，甚是惬意！值得一提的是洗手台采用的透光石材从内部打光，并克服了渗水的问题。 整个销售中心选择使用木材、石材和硅藻泥作为主要材料，旨在打造一种低调而华丽的风格，表现出对法式田园艺术生活的向往。

French Rural Town

The design is a high-end Sales Center. The designer chooses the place where Van Gogh, Cezanne and other artists had long stayed—it is the essence of the design and named Orville which is 30 kilometers northwest of Paris nearby OsieRiver

Orville Sales Center is inclnded two floors. The hall is the whole space which is also the first image. The designer combines painting and relief elements, and integrates them with a specially designed custom chandelier to create a magnificent artistic lifestyle. On the first floor of Orville there is an experience area which is a semi outdoor area with plants and French rural elements. The main materials are plain wooden floors and walls of diatoms mud, with the inside and outside lights combined together to create atmosphere. The second floor is a sequence beginning of the personal space which cannot be too bright. Symmetrical arc-shaped wall is pointed to the plant behind the wine bar. Looking through the French chandelier, passing by the wall, pausing on the plants wall, it reminds of the book you just read. Tasting the French wine, you just feel comfortable and cozy! It is worth mentioning that the washing table made from translucent stone in the interior lighting, and the designer overcomes the problem of seepage. The entire Sales Center's main decorative materials are wood, stone, and diatom mud, and it is made to create a low-key ornate style which is a desire of the French rural life with art.

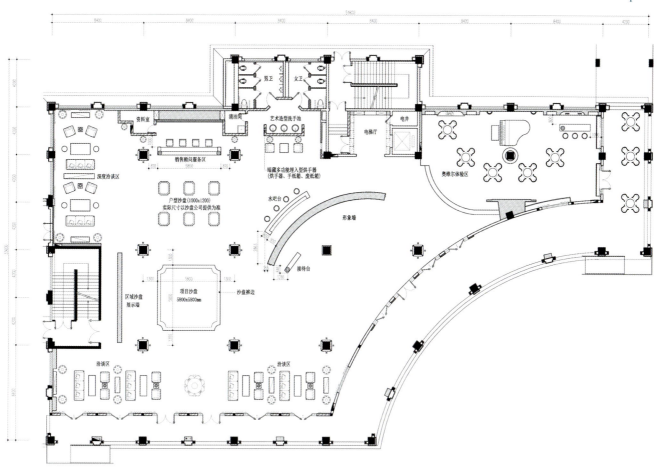

一层平面图

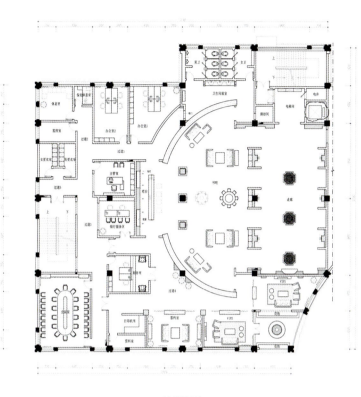

二层平面图

项目名称：盛世领墅
设 计 者：林小真
设计单位：厦门凡城设计工程有限公司

该售楼处的设计以简单、低碳和环保为理念，运用了新型现代材料，在空间上给人以一种奢华的感觉。

开放性的空间布局被规划为模型区、模型展示区、沙盘区、洽谈区、办公区域及休闲水吧区等。各个功能区域既相互关联，又相对独立。销售台背景的一抹红色与洽谈区域椅背颜色相映生辉！

宽敞开阔的空间极具视觉感；大面积通透的玻璃幕墙，不仅弥补了空间光线的不足，还节省了能源。整个售楼空间通体明亮，体现了节能、低碳的概念。把室外的绿化引进室内，让室内添加了一道靓丽的风景。巴洛克式穹顶天花、丝质面料窗帘布艺和镜面雕花玻璃突显了洽谈区域的大气格局。

The sales office was design with the concept of simple, low carbon and environmental protection, using new modern materials. It gives a person with a luxury feeling in space.

The space layout is open and is planning for model area, model exhibits area, sand table area, negotiation zone, office area and recreational water area. Each functional area is connected, and relatively independent. The red background color of sales station and the color of chairs in negotiation zone have embraced.

Spacious space provides us with an extremely visual sense. The large area of glass curtain wall not only make up for the shortcomings of the light in room, but also saves the energy. The whole sales office has a perfect bright space, which reflects energy saving and low carbon concept. The outdoor greening is introduced into indoor room, which is a beautiful scenery for indoor space. Baroque dome smallpox, silk fabric curtain cloth art and mirror carve patterns highlight a general pattern.

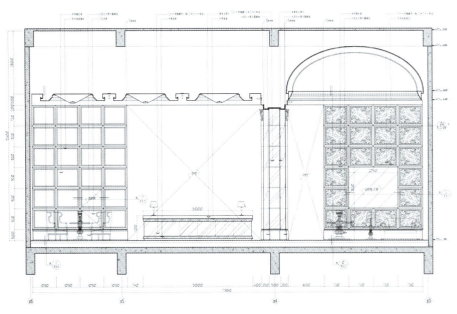

立面图

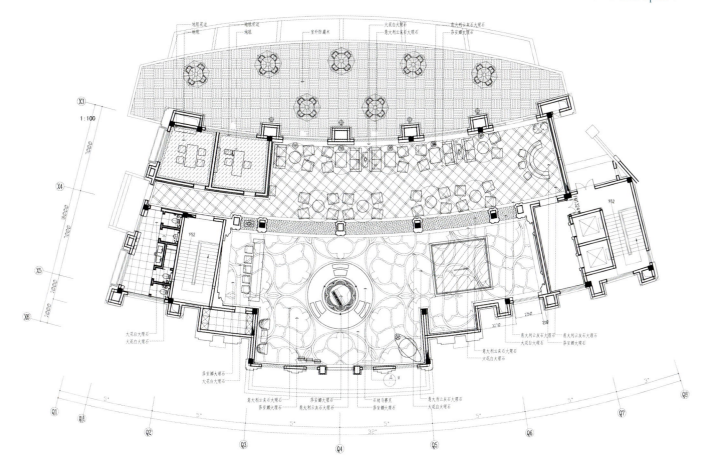

平面布置图

广州市韦格斯杨设计有限公司
GrandGhostCanyon Designers Associates Ltd.

项 目 名 称：四川绵阳东原香屿销售中心
项 目 地 点：四川省 绵阳市
建 筑 面 积：1045 平方米
主 要 材 料：木纹石、西米石、伊丽莎白、雀
眼木、灰镜、银箔框线、镜钢、古
铜不锈钢、墙纸、古铜汽车漆等
陈设艺术服务商：广州市积森装饰设计有限公司
设 计 师：区伟勤、陈晓晖

本项目位于四川省绵阳市游仙经济试验区，由于身处四周环江的地理优势，所以定名为"东原香屿"。为体现其专有价值，售楼部的整体设计风格在强调豪华气派和经典瑰丽的同时亦要贯穿乡水环绕的特色。

在入口模型区，地面采用优雅而大气的大理石欧式拼花，再加上弧形排列而带变化的古铜色不锈钢造型天花板，使整个入口模型区更显得亮丽闪耀，令人眼前一亮。而在洽谈大厅能休闲地眺望优美的江景，天花设计也隐性地呼应这一优势，水纹的柔性天花板反射在弧形的不锈钢造型上，更显气势澎湃。整体空间气氛在营造欧式贵气的同时也带有典雅的时尚感。镜面的合理利用亦令整个空间的气势不断延伸。充分利用设计元素在体现贵气豪华楼盘的同时表达出其优越环境，这样才能呈现给客户一个崭新而向往的生活概念。这就是我们想做的，也是"东原香屿"售楼部要达到的目标。

The project is located in the Youxian economic test area, Mianyang City, Sichuan Province, it is named east of the original sweet island as the geographical advantage of sitting around the river. In order to reflect its proprietary value, the design style of overall sales department emphasizes luxurious and classic beauty and must run through features of the magnificent town surrounded by water at the same time.

The ground uses an elegant and atmospheric European-style marble mosaic and the shape of bronze stainless steel ceiling with changes in arc arrangement in the model area of entrance, so that the whole model area of entrance has shined bright. People can overlooking leisurely the beautiful river in the negotiation hall, the ceiling design also echoes indistinctly the advantage, it is powerful that the flexible ceiling of watermark has reflected in the stainless steel of curved shape. The integral space creates the atmosphere of European extravagance and the elegant fashionable sense. The space has been expanded constantly by making rational use of the surface of the mirror. Making full use of design elements has reflected the extravagance of luxury real estate and expressed its superior environment at the same time, so what we want to do will present a new and wishful life concept for the customer and is also the goal of the sales department of the east of original sweet island.

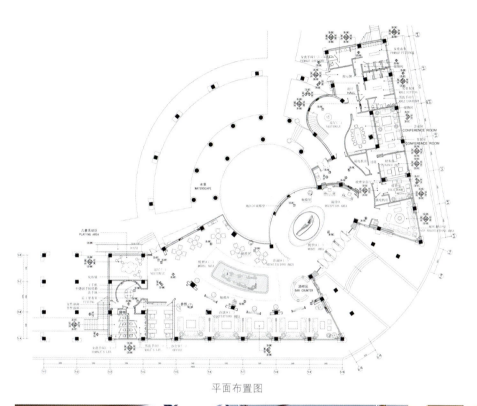

平面布置图

杰出奖 / Excellence Award

①

项目名称: 扬州科派办公家具展厅
设 计 者: 任磊
设计单位: 上海孚若珥建筑装饰设计工程有限公司

扬州科派办公家具展厅设计说明

设计总体概念:"变化的魔方,组合的空间",打破常规布局及构图,采用一致的斜线交叉布局形式。斜角和方形的运用,就像点和线的关系一般,穿插自如,形式丰富,将若干个区域进行交叉组合,犹如办公家具的组合形式一样。展厅不做硬性的隔断和划分,可以进行变化组合。色彩上与企业logo形象色一致,白色为主基调,红与蓝作为点缀色,整个展厅色彩高调统一又有丰富的变化。

可持续性设计理念的体现:

尽量减少吊顶或不吊顶。

采用环保材料。

利用原有空间的结构和梁柱,打造时光隧道,减少工程装饰量。

光源上减少采用射灯、使用LED光源,以节约电力能源损耗。

无中央空调,空调系统主要利用柜机和自然通风来达到调节室内温度的目的。

外建筑采用贴膜的方式达到隔热、保温的功效,减少能量损耗。

组成空间说明:本展厅共有三大部分组成,分别为文化沙龙、时光隧道、旗舰展厅。

文化沙龙:该区域充分利用现有挑高空间的气势,以钻石菱形造型划分空间。主要功能是使客人和参观者有一个可以自由休息洽谈的空间。空间中以斜角形吧台、灯光膜和白色石材吧台的搭配来突出纯净的效果。地面上以沙发和地毯围合成钻石菱形,吊顶与地面呼应,以不锈钢包边和白色灯光膜造型组合成两个钻饰形灯具。

时光隧道:时光隧道则是将企业的发展历史展现在参观者面前,并使用变化的LED光影来衬托出时光的变换。利用原有结构的拱形,加上可以变化色彩的LED灯光隧道,可以变化出丰富多彩的效果,使人流连忘返,充分体现出科派展厅的高科技感。

旗舰展厅:经过了文化沙龙,穿过时光隧道,来到了参观的重点——旗舰展厅。

入口接待区:竖向图案变成了交织的蜂巢形,与时光隧道呼应,象征了科派上下如蜂巢一般的凝聚力和执行力。上方的吊顶上有五个如键盘状的造型,前四个分别贴有桌、屏、椅、橱的英文名称,最后一个为科派的英文名称,说明了科派才是办公家具的最终解决之道。

制造工艺展示区与品牌展示区: 该区域以条纹烤漆玻璃围合成两个C字形,红与蓝两色的对比运用大气而简洁。每个C字形中央都是一个投影设备,分别播放科派的生产制造工艺和品牌阐释,使人对科派的制造品质和品牌内涵有了更深的认识。

新品展示区（或畅销品展示区）： 该区主要突出地面和吊顶的特色。地面采用不同绿色的植绒PVC草坪，看上去犹如天然草坪，衬托出家具的绿色环保理念。

办公类实景展示区1： 该区的概念就是要在展厅做出一个15人的实体公司场景，这也是在实际应用中数量最多的办公空间。完全再现了前台、开敞办公区、总经理室、会议室等，使客户可以有更直观的实景感受，比单纯的展示更具实际效果。

办公类实景展示区2： 从办公类实景展示区1步入展示区2，两个区域功能一样但布局不同。实景展示区2主要体现了办公家具参与装饰的概念。以最少的装饰手法和时间来获得最好的效果，并且可以变化。

精品椅区： 在精品椅区布置了一组对比艺术展台，一个上面放置一张古典的白色明式交椅，另一个则放置了一张贴满了计算机键盘和鼠标的白色办公椅。新旧之间、未来和过去之间的对比充满了趣味，使人可以看到历史的延续和未来的趋势。

材质样板区： 地面以常用的桌面木纹防火板为主，多种色彩相互穿插，延伸到墙面和吊顶侧换成了屏风和墙面上使用的不同布料；色彩变化丰富，这样可以使参观者有一个更为实际的感受。

空中会议区： 从架高的楼梯拾阶而上，来到了全展厅的至高点——空中会议区，在这里可以鸟瞰整个展厅，使参观者可以整体更直观地观看不同的家具组合形态，从而做出更好的决定。

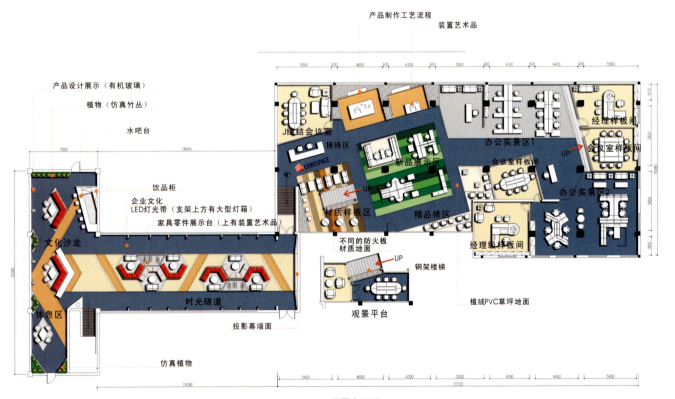

平面布置图

Design specifications for Yangzhou Cubespace Forniture Exhibition

The overall design concept:changes in the cube, the combination of space, break the conventional layout and composition, and use the same slash cross layout. Using the bevel and square,like the link between points and lines,forms and changes. Because havn't high panels and division, several regions can be combined and cross freely. white gine priority to tone,red & blue as the decorative color .The whole color sense in exhibition is unified and in rich changes.

Sustainable design:

To minimize the ceiling or no ceiling.

Use environmentally friendly materials.

Use of existing space and beam structure to create a time tunnel, reducing the amount of construction decoration.

Light source to reduce using spotlights. More LED light source used in order to save electric power consumption.

No central air-conditioning. Air-conditioning system is mainly depend on natural ventilation to achieve the purpose of regulating indoor temperature.

Using the foil on the building outside, in order to heat presenation,insulation and reduce energy loss.

Description of the composition space:this exhibition has three major components— the culture Salon,the

Time Tunnel and the Flagship Showroom.

Culture SalonIn: this area full use of the existing ceiling of momentum space, divided the space by diamond-shape. Main function is provide a rest room. Get the effect of pure through the angle-shaped bar, stone and white light film. Surrounded the synthetic diamond by Sofa and carpet on the ground, ceiling and floor echoed. Stainless steel edging and white light film to form two diamond-shaped lamps.

Time Tunnel: the time Tunnel show the corporate history to the visitors, and use changes in light and shadow to bring out the time of transformation. Using the original structure of the arch, and plus LED lighting tunnel with viriable colors, which can get a variety of effects and fully reflects the high-tech feel of cubespace.

Flagship Showroom:after a cultural salon, through the time tunnel, the focus came to visit - Flagship Showroom.

Entrance reception area:vertical pattern woven into the honeycomb-shaped, echoed with the time tunne, symbol of cellular cohesion and the general

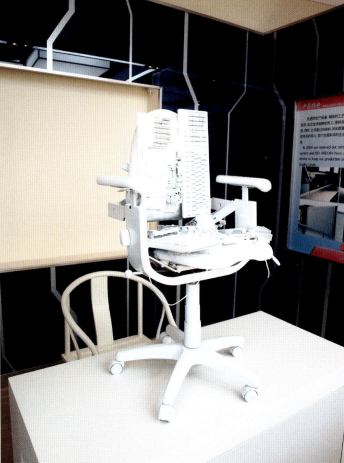

execution of cubespace. There are five keyboard-like shapes on the ceiling above. The first four shapes were attached the words of "table, screen, chair, cabinet" in English, and the last one is "cubespace", which shows us that only cubespace is the final solutio about office furniture.

Manufacturing process and brand display area:in this area around the synthesis of two C-shaped by stripes paint-glass, red and blue and white contrast to the atmosphere and simple to use. There is a projection equipment at each C-shaped central, playing the manufacturing process and the brand of cubepace. It can make people have a deeper understanding about the Division's manufacturing quality and the brand.

New product area: This area will highlight the characteristics of the ground and ceiling. Using green flocking PVC lawn on the ground looking like a natural lawn,and bring out the concept of green furniture.

Real office display area 1:to make an actual entity's 15 scenes in this area.It is also the largest number of practical applications of office space. completely reproduce the reception, open office area, general manager office, meeting room ect, so customers can have a more intuitive feeling than from simply show.

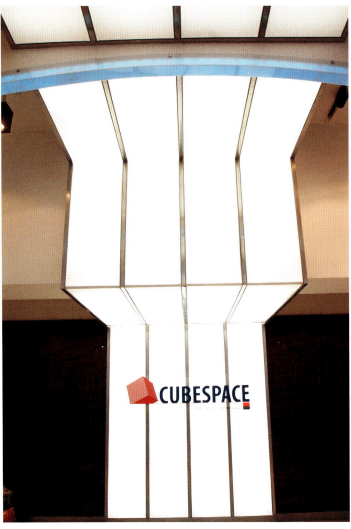

Real office display area 2:from real office display area 1 to area 2, the two regions have the same fuction but diffrent layout. The area 2 mainly reflects the concept of office furniture in decorative:decorative techniques and with minimal time to get the best results, and can change.

Area of chair: a group of contrast art installations. are elaberately arranped in this area.Two stands, one placed above a classic white Ming-style chair, the other placed a white office chair covered with the pattern of a computer keyboard and mouse a white office chair. Between old and new, the contrast between the future and the past is full of fun,and people can see the continuation of history and future trends.

Material model area:mainly using wood fire boardon ground. A variety of colors interspersed with each other,and replaced the different fabrics on the side of panel and celling. Color varied, so that visitors can have a more realistic feel.

Air meeting area: air meeting area is the highest point of the whole exhibition. We can overlook all the exhibition, people can more intuitive view of different combinations of furniture forms, and make better decision.

项目名称：丝绸博物馆
设 计 者：李勇
设计单位：瑞德壹格设计

丝路遗珠

此项目是一个关于丝绸历史的博物馆，它的设计初衷，就是要把巴蜀传统丝绸文化和现代设计风格相融合，在有限的空间内将丝绸文化灿烂多姿的民族风情无限演绎，虽然整个展厅空间比较宽敞通透，但在设计师巧妙的创意布局下变得饱满充实，色彩靓丽的织女做工图版画屏风，造型逼真的中式仿古桑蚕采集间，星罗密布如丝线交错的挑高吊顶，一缕缕红色丝线和木质结构组合而成的传统织丝工具也被架构到了空中，花纹复古神秘的仿古砖拼贴出历史的印记，还有那极富创意的现代工艺为骨、古风神韵为魂的织女造型雕塑，都让这本身只是展示功能的空间变得充满神韵，置身其中仿佛游走在古代丝绸之路的悠悠长卷之中，沿途那些或动或静，或现代或古典的多变风情，都让人沉醉其中。

The Silk Road and the Lost Jewelry

This design is about a Museum of the silk history. Its original idea is aiming at the combination of the traditional silk culture and modern design style. In the limited space, it is showing the colorful ethnic customs of the silk culture. Although the whole space is spacious and transparent, the designer uses ingenious creativity to achieve the full layout. The printing screen with Weaver working on it is bright-colored; the realistic style of the collecting room of silkworm is very antique Chinese. A plume of red silk and wood structure are combined to be the traditional silk weaving tool which is interlaced with high ceiling. Retro- patterned tiles with mysterious antique flavor are the collage of historical marks, and the creative modern technology is severed for the soul of the ancient weaver sculpture. The display function of the space becomes full of verve. It is like walking on the ancient Silk Road, and going along with the changeable style which can be dynamic or static, modern or classical. People are intoxicated in the space.

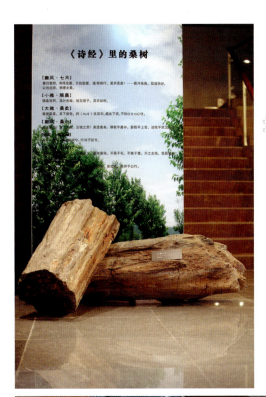

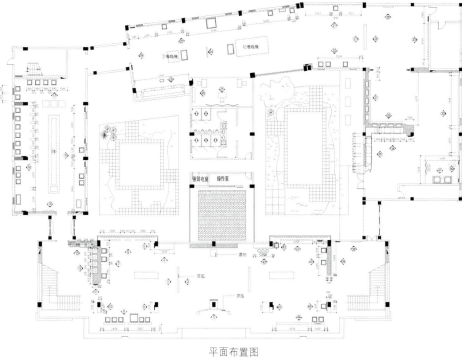

平面布置图

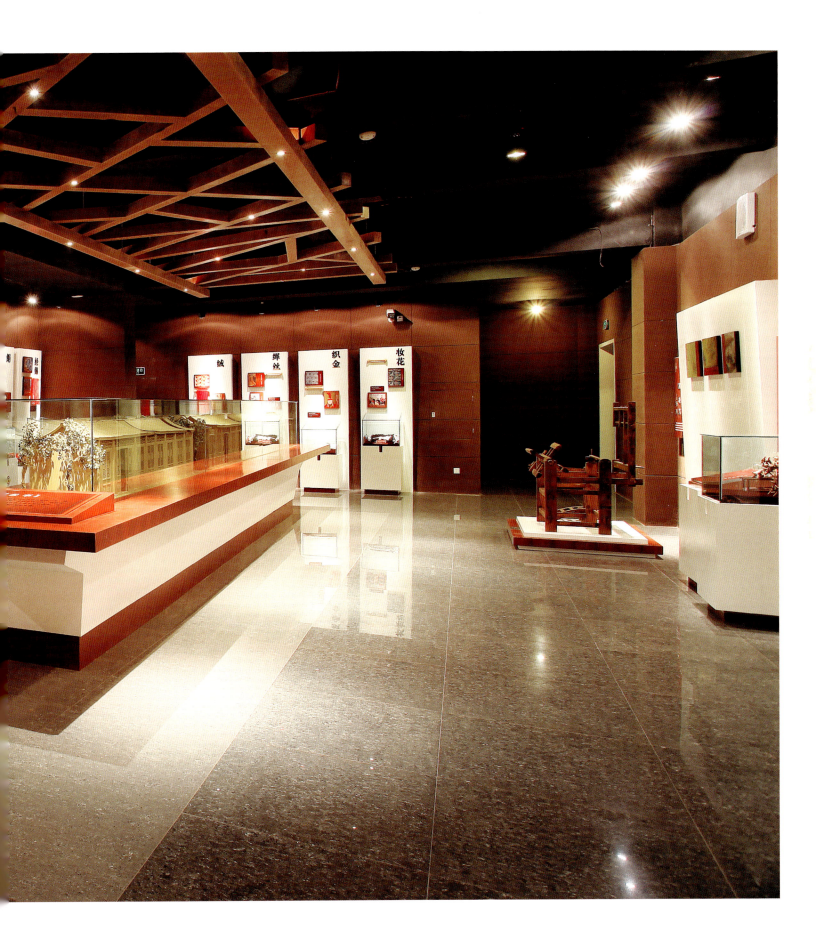

展览空间
Exhibition

银奖 / Silver Award

3

较之室内设计与室内建筑的区别，我认为，重点在于依靠怎样的构成手法进行空间的布置与构成。在此，除了决定以建筑的尺度进行空间的量化设计外，更重要的尝试是，当摒除所有陈设与装饰，它依然是这个样子。

因此，在对光线与供人行走停驻的实体进行组合与设计外，我什么都不做。

To the difference between interior design and interior architect, in my mind, is lies in the skills of the constitution and arrangement of the space. Besides the quantized design by the immersion of the architecture, the more important trial is, when all the decors and displays gone, it remains.

Therefore, I didn't do arything else but combined and designed the light and stops for people.

项目名称：创意亚洲综合楼室内设计——满堂贵金属展
　　　　　示厅
设 计 者：余霖
设计单位：香港东仓集团有限公司

一层纪念品展厅平面布置图

展览空间
Exhibition

银奖 / **Silver Award**
③

项 目 名 称：JNJ mosaic广州马会展示厅
项 目 地 址：广东省广州市天河区珠江新城马
　　　　　　会家居东区
设 计 单 位：大木明威社建筑设计有限公司
　　　　　　(香港/广州/佛山)
主持设计师：谢智明
设 计 师：霍律鸣、叶锦波

作为 JNJ mosaic马赛克广州形象概念展示厅，选址在广州市天河区珠江新城马会家居东区，由于空间的局限性及狭长的店面为设计增加了难度，而在本案处理中，设计师巧妙地运用了蛋卷形时空隧道的空间概念设计，由内而外地打破了狭长而局限的空间限制，同时结合马赛克镶嵌材质的优势及艺术性来表现整个马赛克形象店独特的空间概念及实用性。整个店面分前区的概念空间展示及后区的选样服务区，以合理的空间分隔及混搭的表现手法，展现了一个艺术与功能相结合的马赛克品牌展示厅。

JNJ mosaic image of Guangzhou as a concept show room, located in Tianhe District, Guangzhou Pearl River New City Jockey Club Home Eas.Due to space limitations and the narrow storefront design more difficult, and in the handling of the case, the designers skllfully used the cone-shaped design of the concept of space time tunnel, from the inside out to break the narrow and confined space constraints, combined with the advantages of mosaic materials and performance art to the entire mosaic image of the show room's unique concept of space and practicality. Areas throughout the store falls into two sections: before the concept of space to display and service The designer demonstrated a combination of art and functional mosaic brand show room.

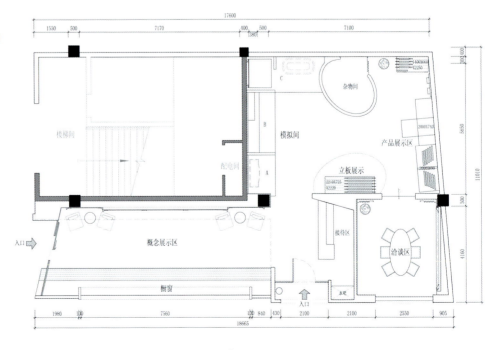

平面布置图

展览空间
Exhibition

优胜奖 / Winning Prize

4

本案设计以人为本，注重绿色环保；功能性与审美性并重。崇尚自然，简约而不简单，追求材质质感。

此项目是由何炅代言的知名品牌E路航产品展厅，地址位于深圳南山区科技园内，由于是公司里面的展厅，前台两边设计了两个开放式接待展示区，主要是展示公司最新，最核心的产品，同时也为接待客户时，让客户第一时间接触到公司最新的产品资讯。

此外，在办公室里面还设计有一个多功能展厅、中柜位置和一个吧台，不经意间就给展厅增添了一些休闲的氛围。旁边还有一个产品体验机，让客户也感觉到产品的模拟使用效果。

在功能和形态上，前台两侧的接待展厅区，则完全是公司形象的展示，中间的白色弧形软膜造型灯下，有一个相呼应的精致的圆形展柜。里面的展厅，两侧的展柜设计成异形，有动感。另外，中岛柜上方顶和展柜运用了产品logo形象色，很自然地把产品文化的理念体现了出来。

项目名称：香港华锋实业有限公司
设 计者：王五平
设计单位：深圳王五平设计机构

Exhibition design

The design is people–oriented, with particular emphasis on environmental protection; both functional and aesthetic elements are not ignored. Advocating natural, simple but not simple, and the pursuit of material texture.

This project is by He Jiong 's well-known brands E road route products exhibition hall, located in Shenzhen Nanshan District science and Technology Park. Because the company is inside the exhibition hall, the front on both sides of the design two open reception display area, and mainly show the latest and most core products. Furthermore for customers, have access to the company's latest product information the first time.

In addition,in the office also has a multi-purpose hall, cabinet position and a bar, inadvertently to the showroom to add some casual atmosphere, there is a product experience machine, and let customers also feel the simmlation of the products use effect.

On the function and form, the front sides of the reception hall area is entirely the company image display. In the middle of the white arc type soft modeling lamp below, have a mutual elaborate circular showcase. Inside the exhibition hall, on both sides of the design into a special shape, dynamically. In addition, the Nakajima above and showcase use the product logo image color, and embodies the products cultural ideas naturolly

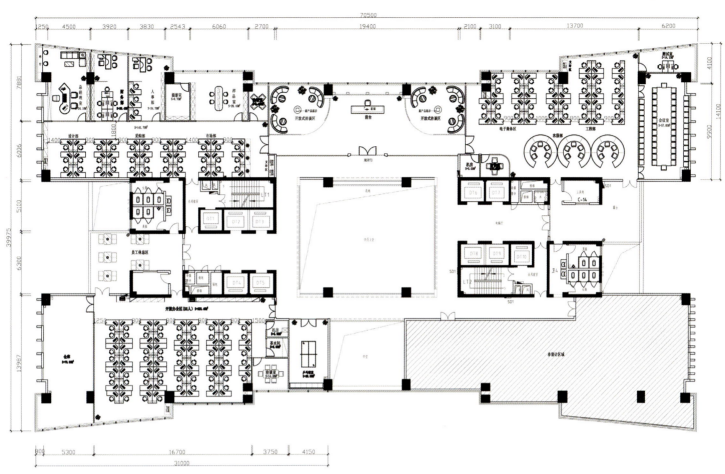

平面布置图

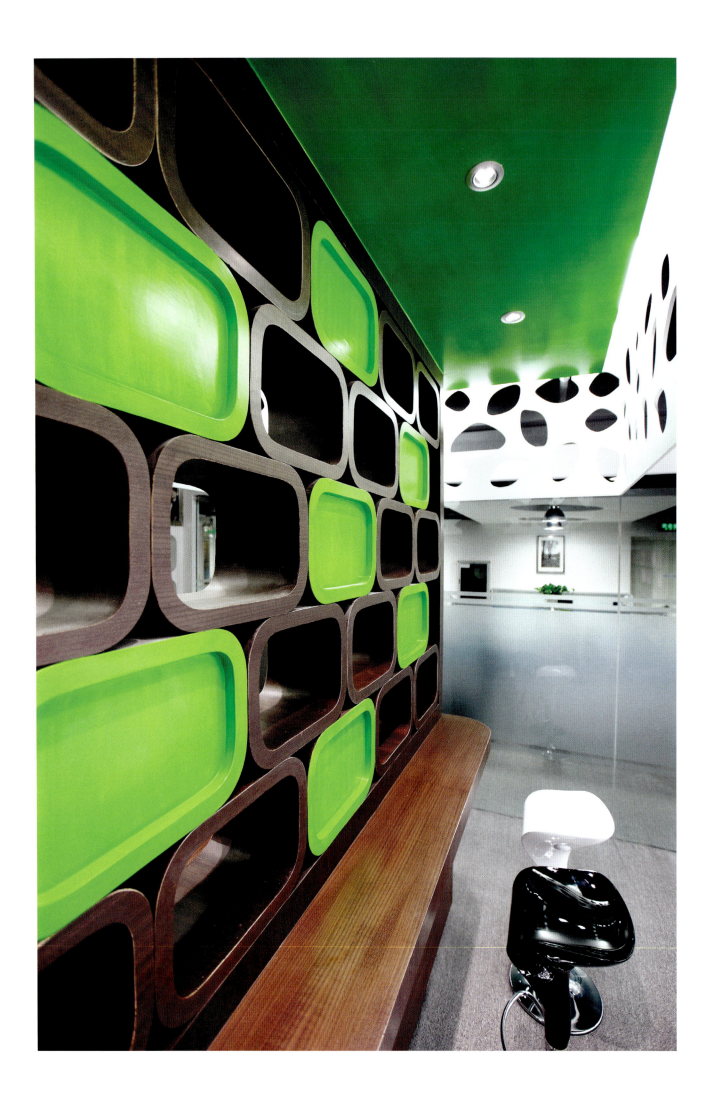

项目名称：春夏秋冬花艺软装配饰中心
设 计 者：梅笑天
设计单位：温州市朗旭酒店设计顾问公司

春夏秋冬花艺软装配饰中心设计说明

春夏秋冬顾名思义就是一年四季的意思。本案是集鲜花、婚庆及软装为一体的陈列空间。

入口处即见涓涓细水长流，卵石间隙处几株植物像个在顽强攀岩的高手牢牢地控制自己的方位。大厅处不经意地挂着几个鸟笼，随着轻音乐飘来阵阵鸟鸣声，迎合了鸟语花香的意境。

视觉的中心是由各种绿色植物组成的一副生动的艺术画，取名为"会呼吸的墙"。坐在菲利浦斯塔克式的吧椅上，一边喝着浓浓的意式咖啡，一边欣赏婚庆的现场片段。

在大厅的某个角落，在休闲沙发上随意地翻阅一下杂志，呷一杯意式咖啡，就可以庸懒地消磨着一下午的时光。

这就是一年四季，春、夏、秋、冬你想要的！

Seasons throughout the year is the meaning of the name suggests – Spring, Summer, Autumn and Winter. This case sets flowers, wedding and software installed as one of the exhibition space.

Form the entrance you will see the trickle, and the gap at the pebbles standing several plants like a rock-climber well controlling his position. Several bird cages hung casually in the hall with the light music and birds singing, which is exactly the same as picture of songs of birds and scent of flowers.

Visual center is an art painting composed of a variety of vivid green plants, named "breathing wall". Sitting on the chair of Philippe Starck's style bar, sipping thick espresso and watching the wedding scene clips SHOW, or in a corner, reading magazines on leisure sofa, sippy a cup of espresso, and you can enjoy a whole lazy afternoon .

This is the year round, spring, summer, autumn and winter you want!

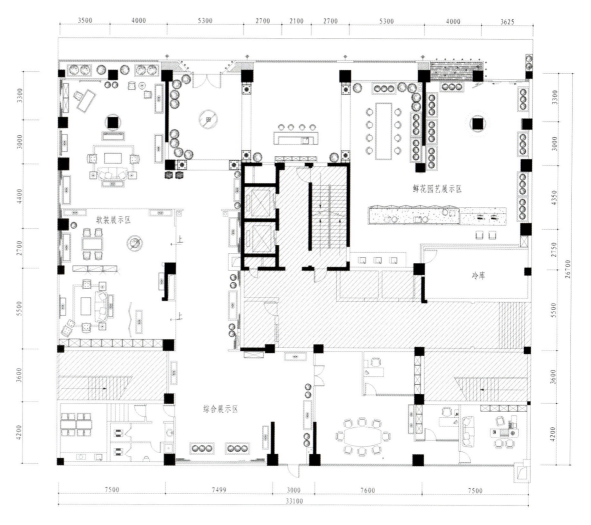

春夏秋冬艺软装配饰中心平面图

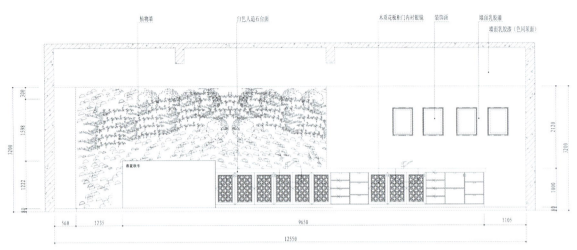

植物墙　　　　　　白色人造石台面　　　　　　　　　木质花板柜门内衬眼镜　装饰画　墙面乳胶漆
　　　　　　　　　　　　　　　　　　　　　　　　　　　　　　　　　　墙面乳胶漆（色同灰面）

操作台主立面图

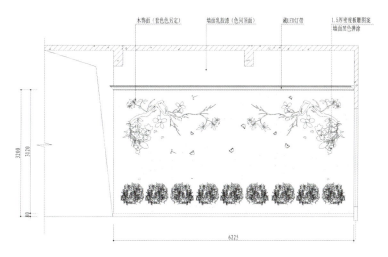

木饰面（套色色另定）　墙面乳胶漆（色同顶面）　藏LED灯带　1.5厚密度板雕图案
墙面黑色弹涂

3200
3120

6225

鲜花园艺展示区主立面图

春夏秋冬
MARRYBRIDES
SINCE 2010

项目名称：首座展厅
设 计 者：蒋晖
设计单位：瑞德壹格设计

风姿卓越

走进这套展示功能的作品，你就会感觉到设计师对各种装饰风格的准确掌控。欧式的活泼、法式的浪漫、中式的质朴韵味，每一个细节都尽显设计师的独运匠心。

沙发宽大、实用、气派、色调柔和，表面可以见到针脚细密有致，用手摇晃沙发也不会晃动。仔细观察休息室的窗帘和摆饰就像是一位位束腰的西洋女郎，风姿卓越。座椅的整个造型也让人喜欢，扶手处设计有灵动的曲线，像深情拥抱的姿势极具人性化。家具的选择上突显人性化设计。在材料的选择上倾向舒适、柔性、温馨的材质组合，可以有效地建立起一种温情暖意的氛围，试想在这样的环境下和朋友喝茶聊天时的惬意舒适。休息室的墙面仍是选择木质材料，配上一副情景写意的装饰油画，显得尤为富贵奢华。

The Remarkable Grace

This design is about to display as a functional work, and you will feel the designer's accurate control of all kinds of decorative styles. Continental liveliness, French romance, and Chinese simple charm, every detail displays the designer's inventive mind.

Sofa is practical and large with soft color, and it has a detailed surface. Even with hand shaking, the sofa is not easy to sway. Carefully observing the curtains and decorations in the lounge, the shape looks like an American girl with the tunic showing her outstanding charm. The shape of the seat is very adorable, and the smart curve of the armrest is designed vividly like a hug full of humanities. The choice of furniture highlights the style of humanization. The designer prefers to the choice of flexible, comfortable and warm materials, which can effectively be established to display a kind of warm and peaceful atmosphere. Just imagine in such an environment you and your friends are drinking and chatting comfortably. The wall in lounge is remained the choice of wood materials. With a traditional Chinese decorative painting on it, it is particularly luxurious.

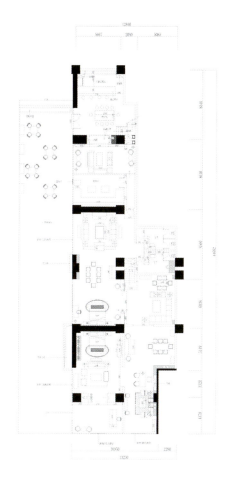
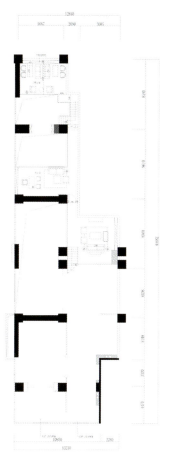

一层平面布置图　　　　二层平面布置图

项目名称：美丽奇缘

设 计 者：Andy Tong

设计单位：Andy Tong Creations Co., Ltd.

美丽奇缘

为营造一个华丽高雅的圣诞氛围并在香港各个商场中脱颖而出，设计师巧妙地将钻石与银河融合起来作为整个装饰的设计概念。设计师特意塑造一颗6米高、14米宽的巨型钻石作为标志性装饰，外观是由不同的镜子组合而成，特显钻石的独特切割手法。除此以外，一棵陪衬着过百种具有钻石形态饰物的8米高的圣诞树更置于巨型钻石之上，增强了整体的圣诞气氛。

设计师在整个装饰上特别利用大量的镜子以增强视觉效果，除了在巨型钻石的表面做处理外，钻石内竟然还有一个由镜子反射出的奇幻世界。首先，客人经过一条由万串闪烁的挂饰及灯光合成的隧道，令客人有如置身于银河一般；接下来客人就会发现此银河隧道是包围着一个由8片单面反光镜合并而成的房间——奇幻森林。此特别设计可以提供隧道空间感及增强视觉效果；沿着隧道，客人亦可从单面反光镜看到另一个世界，令他们兴奋不已。

在奇幻森林里，设计师利用镜子的反射，塑造出一个由8棵圣诞树组成的森林。除此以外，在一个华丽的水晶灯下，一颗1.8米高由透明亚加力及镜面合并而成的钻石，让客户置身于钻石之中。配合由三千颗LED灯组成的十二星座地台图案，浪漫及温馨的气氛弥漫在整个奇幻森林中。

一个全开放式的设计以巨型钻石及其表面镜面的结构处理，安全和巩固性成为首要的考虑，设计师及工程队采用钢铁框架镶嵌了不同尺寸的大型镜面。纵使设计上受到种种限制，整个设计的制作和装嵌过程仅分别在短短3个星期和7天内迅速完成。总成本合计7位数港币。

这是一个香港主要的购物商场，配合以巨型钻石及银河为主题的圣诞装饰及其精巧的表面镜面的处理，特显钻石的切割手法，不但带出了一个华丽、浪漫、温馨及庆祝的气氛，更可大大提高消费者对购物的兴趣。

The Beautiful Edges

To create gorgeous and luxury Christmas atmosphere at one of the main shopping mall in Hong Kong, the concept blending of diamond and the milky way will be incorporated into the festive design.

To become an iconic Christmas decoration compared to other shopping mall, a giant three-dimensional diamond with 6 meters high and 14 meters wide is built. The concept of the giant diamond is referred to the diamond cutting and design. The giant diamond is formed by different size of mirror. In addition, to enhance the Christmas atmosphere, an over 8 meters high Christmas tree is built on the top of the giant diamond. Over hundred of ornaments for Christmas tree are specially designed in different diamond style to incorporate with the theme. To provide visual pleasure, designer wisely using the reflection of the mirror, the purple snowflakes floor patterns are shown on the giant diamond under the reflection. Together with glittering lighting effect, the magnificent, gorgeous and romantic Christmas atmosphere is permeated throughout the shopping mall.

Inside the giant diamond, firstly, the shopper will be shocked by the magical tunnel, which is formed by over ten thousands of hanging chains on the ceiling to create the Milky Way environment. Along

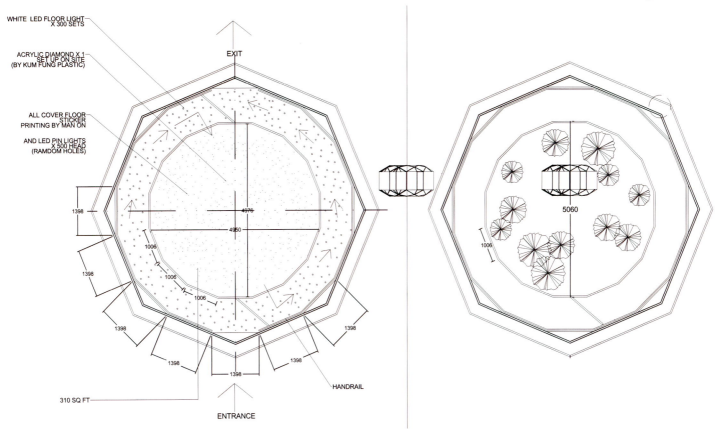

WHITE LED FLOOR LIGHT
X 300 SETS

ACRYLIC DIAMOND X 1
SET UP ON SITE
(BY KUM FUNG PLASTIC)

ALL COVER FLOOR
STICKER
PRINTING BY MAN ON

AND LED PIN LIGHTS
X 500 HEAD
(RAMDOM HOLES)

EXIT

1398
1398
1006
1398
1006
1398
1006
4976
4950
5060
1398
1006
1398
1398
1398

310 SQ FT

HANDRAIL

ENTRANCE

the magical tunnel, shoppers are able to see inside the heart room – Christmas Forest through the one-way mirrors since the heart-room is surrounded by the tunnel. It is not only to provide the spacious visual, but also to let shopper ready to enter Christmas Forest. The heart room is formed 8 pcs one-way mirrors. The Christmas Forest environment is created by the reflection of the mirror which 8 pcs Christmas trees are placed randomly inside the room. Under a classic chandelier, a 1.8 meters high diamond is built for photo-spot purpose. Together with twelve-constellation floor pattern which is formed by 3000pcs LED pores, a romantic, joyful and "bling bling" Christmas atmosphere is created.

In view of the large volume of visitors, the security and sustainability of the giant diamond was a significant concern. Iron steel frame work was erected to support the mirror three-dimensional diamond. Despite of many particular requirements involved, the whole production work was completed within 3 weeks and installment within 7 days.Total costs up to two williow of hkc.

The great location of shopping mall incorporated with the gorgeous diamond design and the impressive Milky Way tunnel, not only to bring a joyful, glittering, unique and luxury environment to the visitors, but also largely enhance the enjoyment of shopping amid the happy and love atmosphere for shoppers.

餐饮空间
Restaurant

① 杰出奖 / Excellence Award

项目名称：厨·聚·酿
设 计 者：Marcos Cain

为精英打造的亚太室内设计奖

此简介从三方面描述烹饪区。厨（THE COOK），具有国际烹饪标准，有12个厨房，在一个非常国际化的氛围中展示烹饪的辉煌。酿（THE BREW），是一个工艺酿造厂，拥有自己的啤酒品牌，由当地酿酒大师所创立，是当地的重点培植项目。聚（THE MEET），设计成为一个现代的牛排餐厅和特色烧烤店，同时营造了一种创新的餐厅和娱乐体验氛围。为了和品牌保持一致，stickman公司把三方面发展成为一个最终的名字：厨·聚·酿（The Cook The Meet The Brew），包括商标、碟子/水杯、制服、菜单、品尝拼盘和打包食品。

一种全新的理念，利用新老材料的创新组合，这样的一种中规中矩的生活主题反映了现代上海餐厅的市场风格，同时又不失国际水准。The cook作为餐厅的一部分和饭店的一部分，里面的可视化展示柜、商品和食物以特殊和交互的视角，让客人体验到当地的氛围。可视化展示台从厨房到桌台创造了一种真正的亲身体验过程。Marcos Cain是Stickman公司的创建人，他设计了一种全新的全天候用餐概念，融合了市场化的现场烹饪和有机组织的餐台，由食物鉴识专家亲自操作。厨（The Cook）内部，有13个食品站，提供广泛的服务形式，每一种均细心达到全球顶尖食物标准，特点包括专门的芝士房子、屠夫肉店、酒窖、咖啡烘烤器和可视化商品展示。

酿（The BREW）是五星级酒店集团Kerry酒店和香格里拉酒店的第一个工艺酿造厂。通过与生产与众不同啤酒的获奖的澳大利亚啤酒酿造厂的共同努力，资本和创造与众不同的餐饮目标结合起来，这是一个新的时代，是全亚洲地区过度复制的餐馆概念。酿造厂技术融合进了餐厅空间。精品酿造厂的设计以产品为导向。我们取消了进口的产品，避免了在当地的生产和二氧化碳的排放。作为200 000平方米的酒店发展的聚合点，注重过程、客人享受互动品尝啤酒的快乐。BREW推出了6款新啤酒和1款苹果汁打入市场与大品牌进行竞争，从而加强自身品质，弥补和定义自己的品牌。自从2011年1月开业以来，获得了亚洲范围的两个奖项，包括亚洲啤酒金奖。

发酵粉用在披萨的底部和啤酒里面从而减少废料产生。Stickman也设计了一款双层玻璃杯，保持啤酒的温度，消耗更少能量，使其恒温。其中一个建议是使用锅炉蒸汽为酒店供热，这样做是因为把热水泵到高层是有困难的。这也是一种有效的为建筑供热的方式。

爱好娱乐、有影响力的当地酒吧就建在酿酒厂周围，是定制的设计和建筑方式。两层楼高的澳大利亚工艺酿酒厂致力于酿造特色口味的啤酒，酒吧以工艺酿酒厂作为可视背景，酒液流经半私密式桌台，上面安装自助啤酒龙头。巨大的啤酒桶成为此地的一部分，看起来很有意思，并且增强了自制酿酒的感觉。在此独特的中心区域融合成功的工程学设计，创造了200 000平方米酒店发展的枢纽区域。酿酒厂现代化的涂饰结合了朴素的可再生木质元素，营造出轻松的风格特点。全新的白色可丽耐大理石插入到大型圆木之中组成了桌子，再利用的背光啤酒瓶堆积起来，在精心造型的圆木之中，创造出装饰墙；打磨的澳大利亚硬币嵌入树脂之中，形成夺人眼目的地板涂饰造型。

在持续性方面，厨·聚·酿所具有的特点导致了一个可持续性建筑，例如，所有项目的生产都是在中国完成的，避免了长途托运过程中的碳排放。再利用的铁路枕木用于建设楼梯。背光机架墙由特殊红砖构成，符

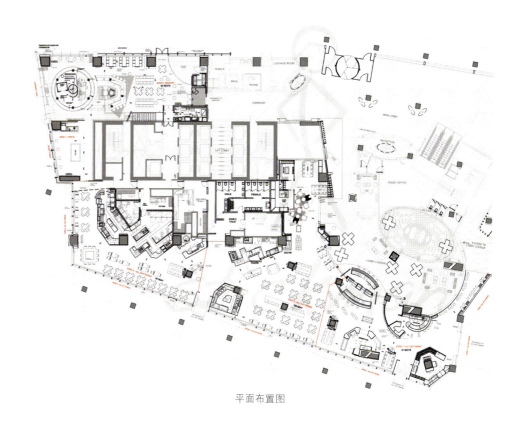

平面布置图

合全天候餐厅和酿酒坊的色调，整体营造一种温暖的感觉。在设计的最初阶段，我们提出了一个椭圆形吧台的概念，用于服务大厅休息室和全部烹饪区，提供实际的服务和能源的解决方案，来减少机器设备的运转，这就需要各级员工在高峰期间加快服务节奏。

照明设备来自于背光再利用的啤酒瓶，位于品酒桌上方。所有特点融合了LED元素。在此区域，我们主推所有可利用的产品，包括蜡纸奶酪包装、各种蔬菜的种子植入包装、环保墨水再生的瓶子用于当地再循环填充。但是，我们没法囊括所有元素。我们具有全球责任感，通过不断努力使得我们的客户从设计中得到欢乐，并拥护可持续性的设计。我们可以骄傲地说，由于在关键区域的精心规划，再利用的砖块、啤酒瓶、硬币地板、木质板和包装和照明的元素的使用一定会有助于可持续设计的发展。

Asia Pacific Interior Design Awards for Elite

The brief was to create one culinary destination with 3 outlets –there is THE COOK; the international culinary experience featuring 12 kitchens in culinary glory displayed as the show case in a very cosmopolitan atmosphere, THE BEW; A craft brewery with its own signature beers created by the resident brew master with the brewery as the focal point. Finally, THE MEET, designed as a contemporary steakhouse and specialty grill, together they create an innovative dining and entertainment experience. To create consistency through out the brand, Stickman Design developed the names of the 3 outlets as one destination: "The Cook The Meet the Brew", as well as the design of the logos, plates/mugs, uniforms, menus, tasting platters and packaging/take-away items.

A brand new concept with an innovative mix of old and new style materials and finishes, all blended together to produce an eclectic lifestyle theme that reflects the deli-style market place of contemporary Shanghai whilst complimenting an international flavor. Within THE COOK the visual display counters, merchandise and food stuffs, allow the guests to experience the venue with a

KEY PLAN ZONE 7 SPECIALTY RESTAURANT

立面图

立面图

unique and interactive perspective as one part deli, one part restaurant. Visual display counters that create a truly engaging experience from produce/kitchen to table. Marcos Cain founder of Stickman designed a fresh all day dining concept combining market-style live cooking with organic deli counters served by specialist connoisseurs of food. Within The COOK, thirteen food stations punctuate the space, offering a wide range of service styles, each meticulously tailored to suit the top quality foodstuffs from around the globe. Features include specialist fromage-packed Cheese Room, meat lined Butchers Block, wine cellar and coffee roaster and visual merchandise displays. The BREW is the first craft brewery in a 5-star hotel group Kerry hotel and Shangri-la group. In a joint effort with an established award winning Australian beer brew house known for an exceptional and distinctive product, the venture would gel with the objective of Food and Beverage differentiation, a new era and approach to Café concepts being overly copied around the Asian region. Brew house technology is to be incorporated within the dining space. The boutique brewery is designed with product in mind. We cut out imported products being beer by producing / manufacturing on site, cutting out CO2 emissions. Serving as a gathering point for a 200,000 square meter hospitality development, with a focus on process, where patrons are interactively encouraged to taste all of the BREW's delights. The BREW has launched 6 new beers and one cider into the market not to compete with the big brands, but enhance there and compliment and define there own brand. Since opening in January 2011, the BREW's beers have won 2 awards within Asia, including the gold medal at the Asia Beer Awards.

The yeast is used in the pizza base as well as the beer thus reducing waste. Stickman also designed a double walled glass to contain the temperature of the beer, consuming less energy

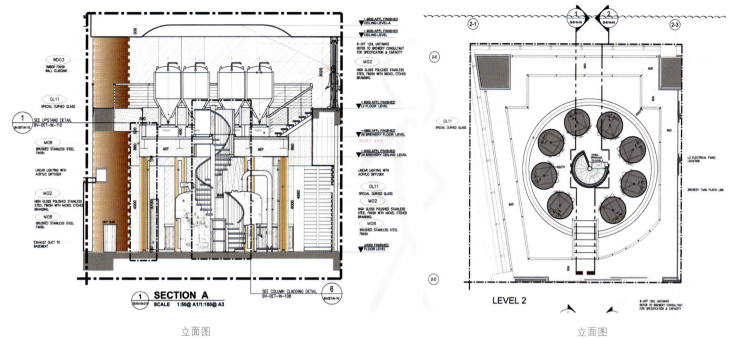

SECTION A
SCALE 1:50@ A1/1:100@ A3

立面图

LEVEL 2

立面图

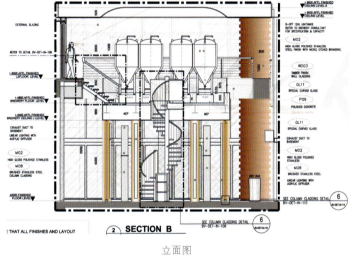

SECTION B

立面图

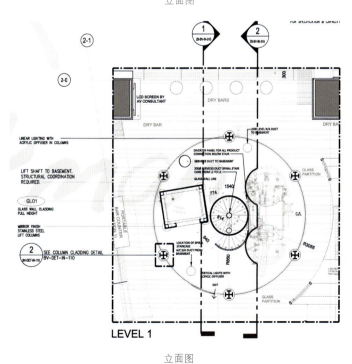

LEVEL 1

立面图

to keep the temperature constant. One proposal by Stickman was to use the steam from the kettle to generate heat for the hotel, because of the difficulty in pumping hot water to the upper levels of a high-rise building; this is an effective way to heat a building. The fun-loving, vibrant local 'pub' revolves around the flying brewery, a custom designed and built, two-tier Australian craft brewery for brewing special in-house flavored beers. With the craft brewery providing a visual backdrop to the bar, the flow through the space passes semi-private tables with self-service beer taps. Huge vats make up a portion of the venue, giving an interesting look and reinforcing the home-brewed feel of the place. A feat of engineering and design combined in this entirely unique centre point create the pivotal anchor of the whole 200,000sq meter hotel development. The Brew's collection of contemporary finishes mixed with rustically reclaimed timber elements creates its relaxed personality.

Fresh white Corian interjects into oversize timber logs to form tables, recycled backlit beer bottles stacked within carefully shaped logs to create decorative wall displays and polished Australian coins set into resin prove to be an eye-catching feature floor finish.

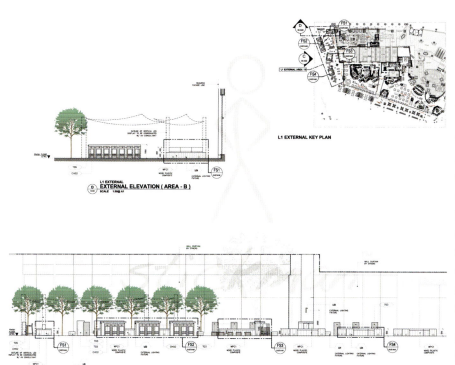

L1 EXTERNAL KEY PLAN

L1 EXTERNAL
EXTERNAL ELEVATION (AREA - B)
SCALE 1:50 A1

剖面图

In terms of sustainability, The Cook The Meet The Brew has elements that contribute to a sustainable build; for example, all production for the project was done in China, avoiding carbon emissions associated with long distance haulage. Recycled railway sleepers were used for the stairs. A feature backlit rick rack wall made from turned red bricks which partially lines the All Day Dining restaurant and brewery creating a glowing and warm hue throughout. In the initial stages of design we proposed a elliptical bar that serviced both the lobby lounge and the entire cook providing a practical service and energy solution for reduced equipment loading this would include staff levels in off peak periods for stocking servicing etc. A bespoke lighting feature made from backlit recycled beer bottles sits over a wine tasting table. All feature/accent/retail lights were incorporated with LED. Finally the within the brand we pushed for all recyclable products including wax paper cheese packaging, seed infused packaging of various vegetables,

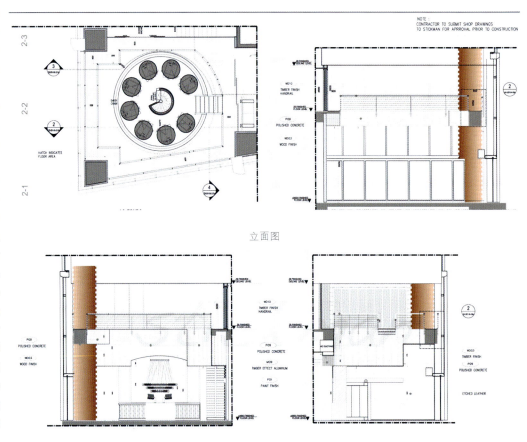

立面图

立面图

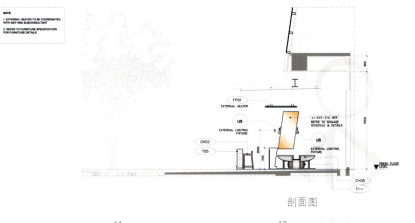

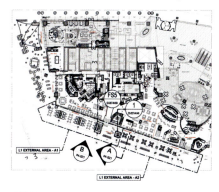

剖面图

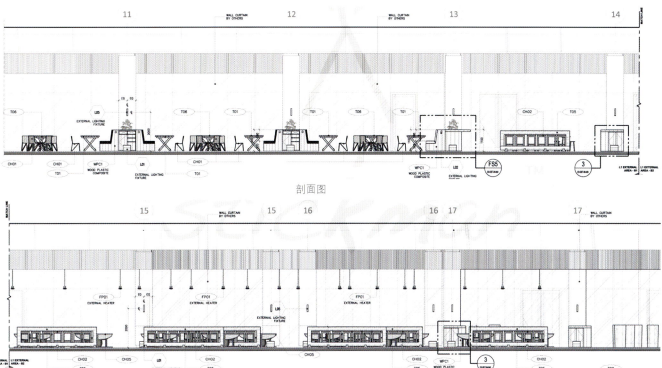

剖面图

剖面图

environmentally friendly inks with a re-usable bottle line for local recycle refills. Unfortunately, we were not able to get all the elements proposed. The point is that we as a design company are actively aware of the environmental impacts with cause and effect and as a global responsibility we constantly push our clients to entertain our ideas and advocacy of sustainable design. Proudly we can say that with careful planning in key areas, the use of recycled bricks, beer bottles, coin flooring, timber panels and elements of the packaging and lighting would definitely assist in the evolution of sustainable design.

餐饮空间
Restaurant

金奖 / Gold Award

项目名称：滩万餐厅
设 计 者：Karen Hay

经过精心设想，坐落于新开业的香格里拉国际酒店附近的四星级餐馆——北京滩万餐厅敞开了木质大门欢迎宾客到来，也成为了标志性建筑。香格里拉集团五星级国际连锁酒店中第八个滩万餐馆已经建立起来，使得滩万餐厅总数目在全球范围达到了33个。

滩万餐厅连锁店拥有两个世纪的优秀传统，深受日本神圣文化的熏陶。自1820年起，餐厅以优雅的招待方式和精致的烹饪料理获得了良好的知名度，并且经常会有皇室成员和世界首脑莅临。滩万餐厅拥有杰出的日本厨师，并自豪于把传统的魅力和现代饮食的进化文化进行不断地完美平衡。

滩万餐厅注重私密空间的保护，大量的私密或半私密独立用材空间为客人们提供了由大厨烹调的精致的日本料理。整个空间包括1个半私密的寿司吧、3个半私密的铁板烧吧、2个私密的点菜区、1个拥有点餐区的贵宾餐厅，总共能提供132个座位。

餐厅不仅能看到北京亮丽的风景，而且用雕塑来诠释日本著名的垂柳树。此项目由Stickman公司和Beau McClellan（照明设计公司）合作，部分是雕塑，部分是照明设计。

大树底部的LED照明灯组合LED天窗板，提供了持久、引人注目的解决方案。以头灯尺寸减到最小，利用垂柳树把光线聚焦在桌子上，桌子由三部分组成，便于场地的交通和组装便利。光源来自于柳树底部的可抽拉板，上面有三个小LED灯，既能聚光又避免刺眼。

大树的每部分都在中国取材并生产。每双筷子都与一个钢铁小环相连接，所以没有必要焊接或使用胶水。高尖端技术把控了周围的光线和中央整体感觉，既轻巧又沉静，既亲密又营造出一种魔幻世界的用餐氛围。

此项目的设计和开发耗费了两年的时间，是Stickman 公司智力的结晶，包括stickman的主管Marcos Cain和其团队，还有北京Nadaman、Beau McClellan的专家，此项目成为了城市的谈资。

After much anticipation, Nadaman, Beijing opened its oversized timber doors to become the signature restaurant crowning the four-level retail and F&B podium adjoining to the newly opened Shangri-La Summit Hotel, Beijing. The 8th Nadaman restaurant for the 5* International Hotel chain, Shangri-La Group, Nadaman Beijing, bringing the total number of Nadaman restaurants to 33 venues worldwide.

The Nadaman chain itself is steeped in two-centuries of rich tradition, and thus all venues are heavily influenced by the sanctity of Japanese culture. Since 1820, the restaurants are known for their gracious hospitality and refined cuisine, and are frequented by royals and world leaders. With leading Japanese chefs the brand prides itself in continually perfecting the balance between the charm of old-world tradition and the ever-evolving culture of modern food.

With a focus on privacy, the numerous private and semi-private exclusive dining areas service guests with fine Japanese cuisine, prepared by a signature head chef. A choice of a semi-private sushi bar, three semi-private Teppanyaki rooms, two private A La Carte areas and a VIP dining room with an A La Carte area, provides a total capacity of 132 seats.

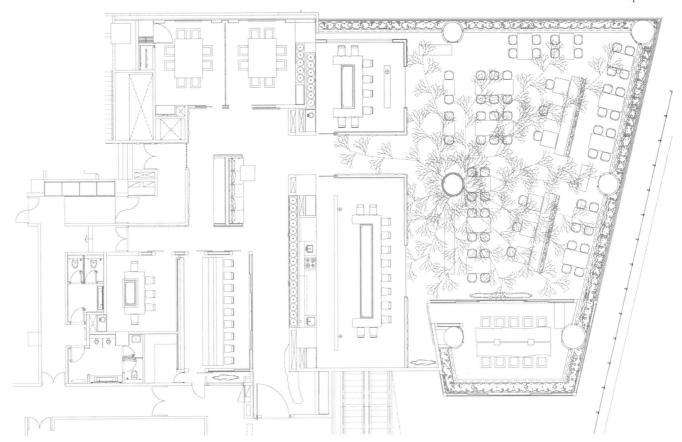

平面布置图

Not only providing stunning views of the surrounding Beijing, the restaurant hosts a central piece d'resistance; the sculptural interpretation of the famous Japanese weeping willow tree. Designed by Stickman in conjunction with Beau McClellan, the famous lighting designer, it is part sculpture, part lighting feature. The 'tree' was hand crafted from over-scaled mirror coated aluminium chopsticks, criss-crossing each other reaching up to the ceiling where the 'branches' span the entire space only to 'weep' stylised droplet pendants over the tables, providing an intimate dining experience. The use of LED lighting at the base of the tree, combined with LED skylight panels, provide a sustainable yet dramatic solution. Overhead lights are reduced to a minimum, focusing the light directly over the personal table settings by long weeping willows which are made in 3 sections for easy transportation and screwed together in situ. The light source on the table is a trio of 3 small LED points mounted on a retractable plate at the base of the willow, focusing the light and avoiding glare.

Each component of the tree was sourced and manufactured within China. The chopsticks are simply connected with small steel grommets so no welding or gluing was necessary. Cutting edge technology enables complete control of the ambient lighting as well as the central feature piece.

Overall the effect is ethereal and calming, intimate and atmospheric all harping back to an era of exclusive dining in a magical world.

Having taken 2 years in the design and development this project is the brain-child of Stickman Director Karen Hay, who together with Director, Marcos Cain and the team at Stickman as well as the expertise of Beau McClellan, Nadaman Beijing is to be the talk of the town.

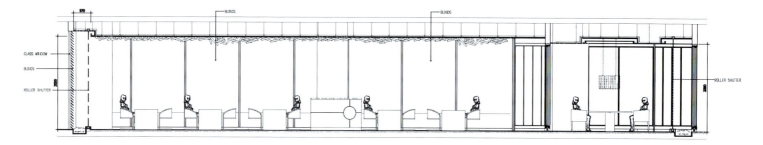

立面图

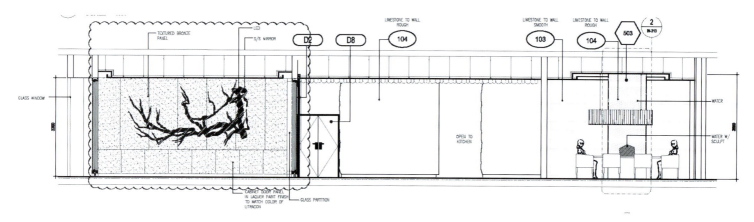

立面图

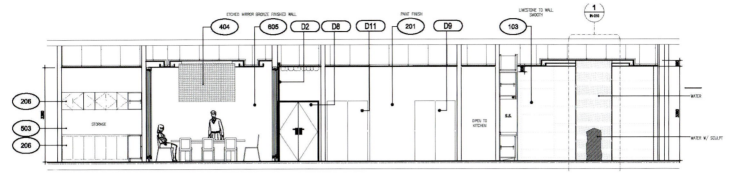

立面图

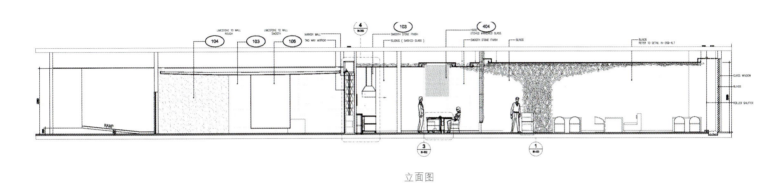

立面图

餐饮空间
Restaurant

银奖 / **Silver Award**

③

项目名称：黄记泰和美食会馆
设 计 师：崔友光
设计单位：光合国际空间设计顾问机构
项目地点：中国济南
建筑面积：800平方米
主要材料：石材、橡木、铜板等

从苏州园林式的入口开始，就注定这是一个不寻常的视觉盛宴。服务台实木根艺玻璃桌，背景是中式米字白色花格，素雅洁净。餐厅走廊的天花板悬挂着白色的鸟笼，灯光略为阴暗，暗色的墙面映衬，墙上的铜花像鱼漂浮在空中，共同成功地营造出一种神秘的感觉。最有意思的是特色大包间里朽木的新用法，将朽木中间对开，把芯的部分抛光作为泡茶桌之用，连椅子都是朽木做成，时光的足迹雕刻在木头上，也雕刻在我们心中。鱼浮游在圆圈意向的水的波澜中，禅味十足。素色水泥墙上的木头太阳花造型，黑色的花透出斑驳的白，人文色彩浓郁，粗砺的原木配上精致的圈椅，一样的耐人寻味。餐区的木隔断颇有东南亚风格。

Suzhou garden style entrance from the beginning, it is destined to an unusual visual feast. For the root carving wood desk glass table, the background is white Chinese rice word lattice, clean and elegant. Restaurant corridor ceiling white bird cage, lighting, slightly dark, dark against the background of the wall, the wall of the copper flower floating in the air like a fish, the success of creating a common sense of mystery. The most interesting feature ars large rooms, new uses of dead wood, dead wood off the middle, the core part of the polishing tables for using as a tea, and even the chairs are made of rotten wood, carved in wood time footprint, but also sculpture in our hearts.

Fish float in a circle of water waves in intention, full of Zen. Sunflowers plain cement wall wood shape, revealing a black mottled white flowers, cultural rich color, of crude coupled with exquisite wood chair, the same thought-provoking. Wooden dining area is quite cut off Southeast Asia style.

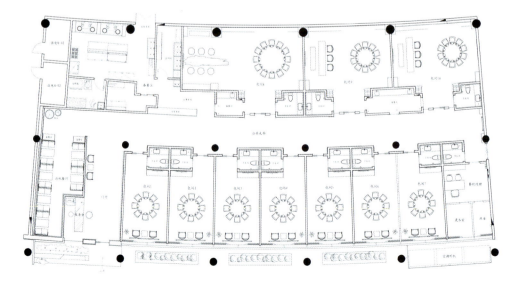

和瑞园会所平面布置图

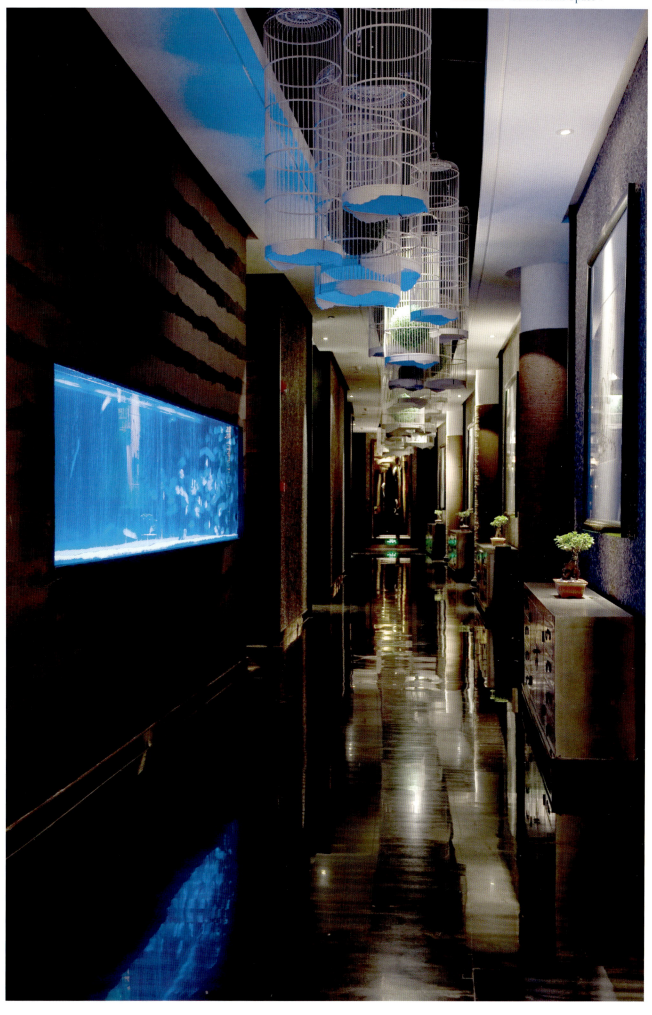

项目名称：福州赤坂日本料理餐厅
设计公司：品川设计
设 计 师：郭继
面　　积：350平方米
主要用材：黑钛、水曲柳木饰面、砂岩、瓷砖、金刚
　　　　　板、青石

流淌的日式气息
福州赤坂日本料理餐厅

在日式风格的设计理念中，设计师一般比较注重体现浓郁的日本民族特色，在选料上注重质感的自然与舒适，常选择木格拉门、地台等元素来表现其风格的特征。在本案设计中，设计师除了保持日式餐厅的传统风格外，更增添了一些现代时尚的符号。餐厅内部环境雅致且轻松简约，布局独具匠心，散发着东方古朴的异域风情。主要材料为日式中常见的木制材料，将不同颜色质感的木制材料相融合，配合砂岩、金刚板、青石等比较刚毅的元素，其刚柔并济的视觉效果为我们带来幽雅舒适的就餐环境。在布局上，设计师运用完美的衔接与技巧，使得每个区域都功能明确，却没有硬性的区分界限，一切显得那么自然融洽，吸引着人们去细细探究。

惊艳不需要绝对的夸张，细节的完备才可以成就个性的完美。在这个350平方米的空间里，设计师将时尚与传统的符号相互融合，巧妙地运用到每一个环节，趣味和创意如舞动的精灵般吸引着人们的眼球。餐厅里随处可见的浪漫樱花、精致考究的的吊顶与吊灯、流淌着灵动气息的水池与鹅卵石、带有浓郁日式风格的屏风与拉门，还有那些造型优美且高雅舒适的餐椅，都是餐厅最为吸引人的亮点，体现了设计师的独具匠心。

无论是身处时尚典雅的大厅，还是围坐在日式风情的榻榻米包房，置身在这样一个曼妙的环境中，触摸它动人的情调，略显神秘的气氛，在这个轻松流动的空间，感受着一种高品质的生活，那是一种格调，同样也是一种人生。

Japanese Cuisine of Akasaka, Fuzhou

In the Japanese-style design concept, designers always emphasize the strong Japanese features and are partial to the natural and comfortable materials. They always choose sliding doors with wooden lattice and platforms to display the features of the style. In the design of this project, the designer continues the traditional style of the Japanese restaurant and increases some modern and fashionable symbols to make the restaurant become a relaxing and delicate space, helping customers experience the elegant and comfortable concise living style.

The interior environment of the restaurant is elegant but concise and relaxing with unique layout and pristine exotic flavor of the orients. The main material is the common wooden material in Japanese restaurant, with combines different kinds of wood with different colors and textures, together with hard elements like adamas plate and bluestone. The visual effect brought by combination of hard and soft elements produces an elegant and comfortable dining environment. In layout, the designer adopts perfect skills to define the functions of different areas but without rigid boundary. Every thing presents to be natural and compatible, attracting people to explore carefully.

Gorgeousness needs no absolute exaggeration. Perfect details would fulfill perfect personality.in the space with a size of 350☐,the designer combines symbols of fashion and tradition and skillfully applies

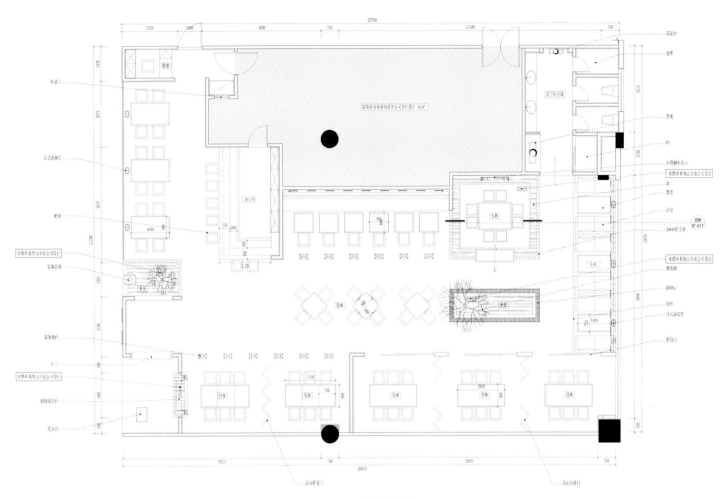

平面布置图

them to every detail. The interest and creativity in these sybols attract people a lot.

The romantic sakura prevailing in the restaurant, the delicate and sophisticated suspended ceiling and ceiling lamps, the smart pond and cobble stone, the screen and sliding door in stong Japanese style, and the elegant and comfortable dining chairs with beautiful shapes are the highlighting points of the restaurant, revealing he designer's creativity.

No matter it if the fashionable and elegant hall, or it is in the tatami compartment in Japanese style, you can put yourself in the graceful environment, feeling its touching sentiment and mysterious atmosphere. In the flowing and relaxing delicate space, you could enjoy a high quality life which is a kind of living style and also a kind of life.

可持续　舒适空间 | **151**
Sustainable Comfortable Space

项目名称：碧翠茶楼
设 计 者：杨洋
设计单位：成都斯韦普（香港）设计顾问有限公司

竹情画意

很多茶客品茶往往品的是一种心境。一个静谧舒适的环境配上一壶淡雅的清茶，生活的情趣就随着那上浮的茶叶渐渐满溢。这套作品是一套古色古香的中式茶庄，设计师在构思的过程中充分利用中国古典元素，将营造茶庄的氛围作为本次设计的重点，可谓用心良苦。一个翠竹清风、诗情画意的古典空间呼之欲出。

设计师深知爱茶之人对竹子也有天生的喜爱，所以就别出心裁地用交错的竹排做空间顶棚，让整个空间充满室外的自然气息。垂钓在空中的卷轴犹如幔帐增添了人文气息，墙壁则是青青石板，悠悠子衿，与竹韵顾盼生辉。最让人侧目的是那和天地相连的古代美女屏风，飘逸的神韵让人不禁恍惚，今昔何年都已成浮云，唯有这竹情、画意、美眷、清茶才是真正的人生。

Emotion of Bamboo and Painting

Many people like to have a tea in a cheerful frame of mind because quiet atmosphere and a cup of mild tea make life leisurely. This work is a traditional Chinese tea house, and the classical culture reflection is the core principles of design to achieve a beautiful and harmonious classical atmosphere.

Designer knows that people, who love tea, would inevitably love bamboos as well. Correspondingly, the ceiling is covered by crossing bamboos raft so that indoor space looks like natural wild scene. The hanging scroll, painting on the wall and the traditional beautiful lady screen are intoxicating so as to be self-forgotten but only bamboo, painting and beautiful scenes consist of a real life.

平面布置图

项目名称：福州廊桥会所

项目面积：600 平方米

项目地点：福州

设计单位：福州多维装饰工程设计有限公司

设 计 师：林洲

主要材料：海棠瓷砖，金刚板，壁纸，地毯，金箔
漆，英国棕石材，文化石，灰镜

供稿单位：福州多维装饰工程设计有限公司

梦临廊桥

每一座廊桥都有一段优雅的故事。在夕阳缭绕的午后，蝴蝶穿插飞舞，桥头重逢的恋人和清冽潺潺的流水，使人们想起廊桥的美好印记——在弗朗西斯卡和金凯的廊桥遗梦里，在白娘子和许仙的断桥相逢中。而廊桥会，也是一个有故事的地方。

廊桥梦境

当你步入廊桥会所，一汪碧水荡开,奠定了整个空间的基调。碧绿的青草和盛放的夏莲，如同徐徐伸展的画卷；巧妙隐藏的光源，氤氲出一片柔美。通道空间宛若一处如痴如醉的廊桥梦境。尖顶的天花板设计吸纳了廊桥的经典元素，古朴斑驳的石头墙面无声地刻画着旧时光的优雅传承，富有纹理的木地板如同桥面延伸，在静默中曲折缭绕，沉淀着深沉而浓厚的人文气息。

轻扬质朴

三三两两的石柱体零落桥畔，精妙的各种水生动物雕饰栩栩如生，如同历史的守护者，沉淀着别样的美感。而在包厢区域与外界的衔接上，以钢构玻璃为隔断，营造了一个自由通透的就餐环境。轻曼的珠帘犹如席席倾洒的水珠，律动着廊桥的美景。垂吊的花灯造型感十足，将廊桥的古典神韵延伸至包厢内部，墙面是"接天莲叶无穷碧"的大朵莲花，地面则是柔和的仿古瓷砖，各种色彩在此相互巧妙过渡且完美融合，一种轻扬质朴的氛围在空间轻柔地流动。

飘渺之境

拾级而上，大块的玻璃镜面映射着两端风景，二楼大堂之上，一幅巨大的山水画作成为空间的主角，画中云雾渺渺，群山相互呼应，灵秀动人，流露着高雅的生活品位。明式家具造型简约流畅，两扇婉约的灯笼屏风伫立两侧，升华了空间的传统人文意境，整个氛围宁静而致远，美不胜收。

Chinese Lounge Bridge Restaurant, Fuzhou

There is always a story behind one place, such as a heart-touching story for The Bridges of Madison County. And Chinese Lounge Bridge Restaurant is no exception.

A pond of clear water welcomes guests upon their arrival and set the tone for the whole interior space. Green grass and lotus in blossom look like an unfolding picture in blurred light by hidden source. Dreamy corridor, pointed ceiling take in classic elements of Chinese lounge bridge. Rustic and mottled stone walls quietly seems to have stories in old days to tell and textual wood flooring winding through the entire restaurant carries profound cultural quality.

Stone pillars scatted next to the bridge and vivid carvings of aquatic animals reveal special beauty. Toughened glass is used as partitions between public area and private rooms and crystal clear dinning environment enhanced by graceful bead curtains which look like dropping water drops. Tasteful hanging lanterns bring the rustic charm of Chinese lounge bridge into the private rooms which are clad with huge lotus-patterned walls and paved with gazed tiles. The space boasts of smooth transition in color and is

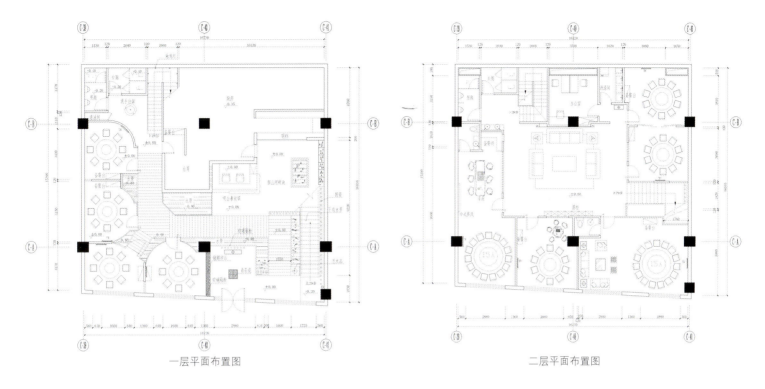

一层平面布置图　　　　　　　　　　二层平面布置图

overflowing with lovely rustic feel.

Going up by a staircase with large glass on its both sides to the second floor, a huge landscape painting is found in the focus. The painting of mountains covered with fug and clouds reveals a good taste. Neat and simple furniture of Ming Dynasty style with two lantern screens on its both sides add cultural taste to the space and develop quiet grace.

优胜奖 / **Winning Prize**

4

项 目 名 称：贵州铜仁上座会馆
面　　　积：7000 平方米
项目设计时间：2010年1月
项目完工时间：2011年3月
所 用 材 料：大理石，绿可木，布艺沙发，装饰画
设 计 单 位：深圳华空间机构
设 计 师：熊华阳

坐落于贵州省铜仁市的上座会馆，是接待当地高端人群及高级客商的重要场所。对于这样一座有山有水的自然静寂之城，最适合于它的建筑莫过于传承我们传统文化的中式风格建筑了。上座会馆由两座三层高的中式建筑砌合而成，从建筑外观到室内设计，及软装配饰等，均由我们一体化完成。会所内含有健身会所、娱乐酒吧、中式餐饮等项目。

对于设计的前期，需要设计师与甲方沟通此项目的市场定位及各方面要求，并结合当地的消费市场做调查与研究。根据调研结果，我们先分析出其目标客户的生活喜好、欣赏品位、消费习惯等，并结合当地的经济发展状况及未来的城市发展目标给会馆做长期的发展计划，再设计出一系列的建筑设计及室内设计方案，世外桃源般的上座会馆由此产生。

设计师在设计项目时需扬长避短，充分利用项目的优势。上座会馆依托于当地的山水之景及少数民族特色，从外观设计到室内设计，都使用中式的设计框架，并结合现代风格的家具、饰品，院子中央的小池塘、少数民族特色的壁画、现代风格的沙发、中式古典的木椅、竹叶图案的地毯……使会所由内及外散发出新中式设计风格的高雅。

空间的设计不在于将墙面及吊顶做复杂的处理，而在于给它恰到好处的点缀；为了使古典中式的沙发不过于呆板，我们结合现代简约的沙发一同陈设，流畅的线条即显现出来；为了使普通的过道拥有设计感，我们使用带有图案和镜面的玻璃做墙面；为了使正统的中式包房不止拥有古典的中式家具，我们设计了与众不同的玄关增加了包房的设计感；为了不忽视楼梯的设计，但我们注重项目的每一处细节，选用几何形状的时尚楼梯扶手，并在楼道陈列着艺术品展示。

无论是室内设计，还是产品设计，成功的设计在于相辅相成，由点及面的互相呼应，正如此案中多处应用的方圆结合。会馆外观、接待大厅、池塘边上、包房内的玄关，都是方圆之间的艺术组合。设计师在建筑与池塘之间留有数平米的空间，并设计成临水而坐的闲叙之地，凉爽的微风、静谧的湖面、舒畅的空间，非常受宾客欢迎。

在装饰材料的选用和软装的陈设上，设计师也花费了诸多的心思考察市场。美观且耐用的家具，古典且时尚的艺术品，室内装饰材料的颜色搭配，家具的设计，艺术品的摆放角度，绿色植物的装饰，以及使用字体的选择……总之，项目的设计不仅在其"面"，更重要体现在其"点"上，在其一点一面的设计与搭配之间设计中决定着项目的成功。

Shangzuo Club House in Tongren, Guizhou, is the important place to receive local high-end people and high-grade customers. In the quite and natural city with hills and waters, no building is more suitable than the Chinese style building with thousands of historical culture. Shangzuo Club House is bonded with two blocks of three-floor Chinese buildings including fitness club, entertainment bar, and Chinese catering. The Hua Space Design Institution has completed the integration of all aspects including appearance of building, interior design and soft assembling and decorations.

In the earlier stage of design, the designers have fully communicated with Party A about the market

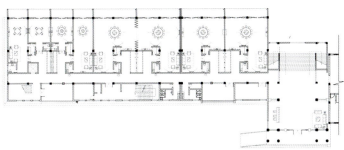

A栋一层平面布置图

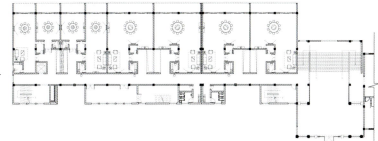

A栋二层平面布置图

positioning and requirements on various aspects of the project, and performed investigation and research by combining with local consumption market. According to the survey result, the designers have firstly analyzed the life hobbies, appreciation grade and consumption habit of target customers, and made long term development plan according to local economic development status and future city development goal. Then the designers have made a series of design proposals on construction design and interior design.

The designers shall adopt their good points and avoid their shortcomings during the design process to make full use of the advantages of the project. Based on local scene of hills and waters and features of minorities, Shangzuo Club House has applied Chinese design framework to combine with furniture and decorations of modern style, small pool in the center of yard, mural with features of minorities, sofa of modern style, wooden chair of classic Chinese style, and carpet with bamboo leaf in the appearance design and interior design. As a result, the Club House has expressed traditional and elegant new Chinese design style from the interior to the appearance.

The design pith of space lies in the suitable ornament instead of complex design on wall space and suspended ceiling. As the classic Chinese sofa may be certain rigid, so the proposal has located the classic Chinese sofa together with modern and simple sofa, which has reflected the smooth line at once. The wall face is made of glass with mirror surface and pattern, so the common corridor has also reflected the sense of design.

A栋三层平面布置图

Both for interior design and product design, the successful design lies in complementary and mutual cooperation from the point to the face. For example, the proposal has applied the combination of square and circle for many times. The artistic combination of square and circle has been in the appearance of club house, reception hall, surrounding of pool, and hallway in compartment. Designers have left space of several square meters between the building and the pool to design a place for chatting in front of wind. With cool wind and quite lake face, the happy space has been very popular with customers.

On the aspects including the selection of decoration materials and the location of soft decorations, the designers have also performed careful survey in the market. In a word, besides the "face", the design of the project has paid more attention the points. The success of the project is determined by the design of details and the design and collocation of the face and the point, such as beautiful and durable furniture, works of art with classic and fashionable style, the color assortment of interior decoration materials, design of furniture, location angle of artistic works, decorations of green plants, and selection of script...

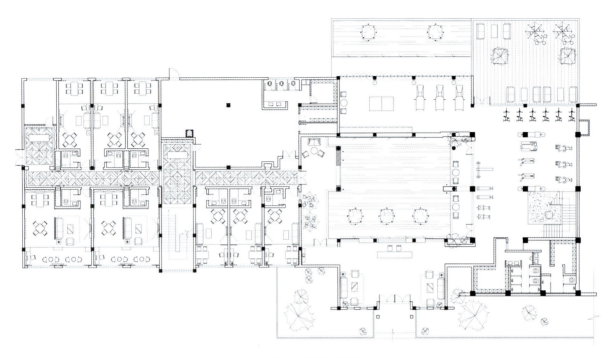

B栋一层平面布置图

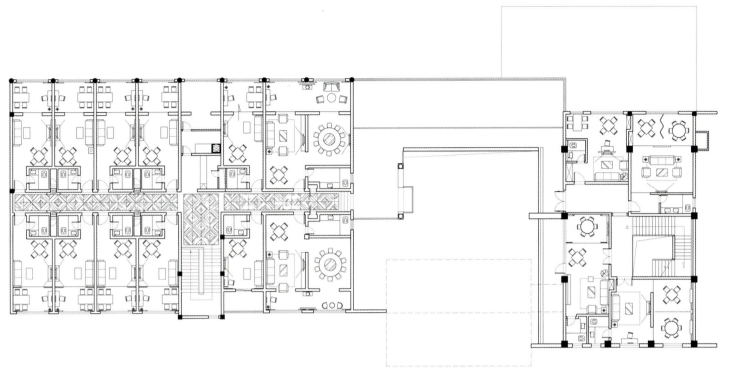

B栋二层平面布置图

项目名称：览月汇
项目地址：万达三层
设计机构：福州品川装饰设计有限公司
主要材料：金刚板、玻化砖、大理石、玻璃LED、
真石漆
设计说明参考：现代风格

时尚餐厅的浪漫主义色彩

览月汇位于万达广场三楼。当人们从现实的喧闹中走进这个暗色调的空间里时，那份浮躁便慢慢褪去，内心的平静带来的满足感充盈了整个肢体。设计师在灯光与环境色彩的营造、景观的运用和材料间的对比协调中让空间产生了丰富的戏剧性。

在这个融合了现代简约与梦幻元素的空间里，设计师将唯美的浪漫主义融入现代的风格中，以稳重、精致的色调及充满奇幻色彩的光线为主轴贯穿整个空间。在餐厅中央有一列由玻璃LED组成的"格子间"，将整个就餐空间大致规划为三列。"玻璃格子"的尽头是一弯月牙的造型，LED灯不时变幻出蓝、红、紫等色彩，在各种梦幻色彩的映衬下，浪漫情怀被演绎得淋漓尽致。在此用餐时如同飘浮在夜空中，伸手便可揽月。在这样的空间里，无需强烈的灯光，恰到好处的昏暗光线反而容易令人产生遐想，为这里增添无限神秘的色彩。

设计师将金刚板、玻化砖、大理石、玻璃、真石漆等材质巧妙地融合应用，使得空间的表现力得以扩张，其独特的搭配效果也让装饰成为空间的视觉主体。同时，设计师运用完美的衔接与技巧，将每个区域的功能明确表达出来，却没有做出硬性的区分界限。一切显得那么自然融洽，吸引着人们去细细探究。

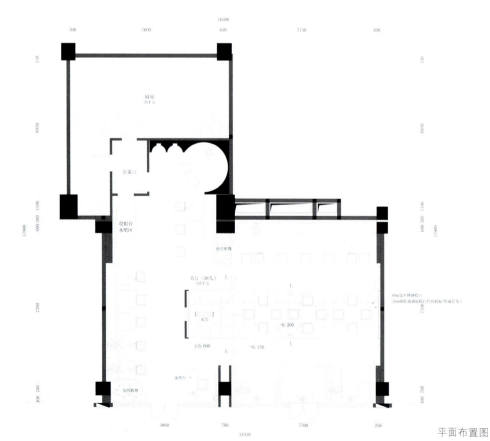

平面布置图

Romanticism of the Trendy Eatery

Royal Will locates at the third floor of Wanda Plaza. When you come to the space in dark hue from the noisy reality, the fickle heart will feel calm gradually, and the peace in heart will bring you comfort and satisfaction. By the creation of light and interior color, application of sight, and comparison and harmony of material, the designer makes the space abundant drama.

In this space integrated with modern simple and fantastic elements, the designer integrates the esthetical romanticism into modern style; it takes the steady and exquisite hue and the fantasy magic light as the principle axis to cross the whole space. There are some "grid room" made of power glass in the center of the dining hall, and it divides the whole dining room into three lines. At the end of the "power glass grid room", there is a crescent sculpt, which performs the romance clearly against various fantastic color by the LED light. Having dinner here is just like floating at the dark sky, which makes you think that the moon can be fetched as you will. In the space like this, no strong light should be used. In contrary, proper dark light is good for imagination, and adds limitless mystery.

For the application of material, the designer integrates the emery board, vitrified tile, marble, power glass, and stone painting to enlarge the expressiveness of the space. The unique effect of the collocation also lets the decoration become the main vision of the whole space. At the same time, by the excellent skill, the designer expresses the function of each area clearly but not harshly. Everything here seems nature, which draws people's attention to explore.

项目名称：上海石榴酒吧
设　　计：王善祥
参与设计：王善辉、龚双艳、张笃
摄　　影：老猫
设计时间：2008年2月～2010年3月
完工时间：2011年3月
面　　积：230平方米
主要材质：耐火砖、水泥抹灰、旧木地板、青花小瓷
　　　　　砖、旧家具等

这是一个边设计、边施工、边营业的项目。项目从一开始就不停地改设计、改施工、改营业定位和方式，同时业主合伙人也一直在更换。

石榴酒吧是一个230平方米的呈8字形状的空间。朝向马路的外侧略小些，内部稍大些，但没有窗户。对面是个公园，白天景观不错。酒吧白天至傍晚提供饮料以及具有西域风味的（西式）食品，并有原创乐队现场演出。主要客户群是以在上海的欧美人和喜欢现代休闲娱乐的中国人为主。

相传石榴是西汉张骞出使西域时引进中国内地的。西域，包括今天的新疆维吾尔自治区以及阿富汗等中亚地区在内，是个多民族文化融合的地区。

投资项目固然是以赢利为主要目的，但赢利的方式却和投资者的文化背景密切相关。石榴酒吧的几个投资者有维吾尔族、回族、汉族。"Anar"是店铺的英文名字，是维吾尔语"石榴"的发音，朗朗上口。投资者希望店铺既具有国际风格，即俗话说的"洋气"，同时又要融入西域风情，体现出多元文化的融合特征，以特色立足市场。

酒吧做得"洋气"是不用费劲的，因为其本身就是西洋事物。西域风情是什么样子，恐怕大多数人都难一下子说清。很多人都知道，广袤的沙漠、戈壁草原、雪山、新疆民族歌舞等等这些无一不是西域风情。可是如何在一个230平方米的小空间里融入西域风情，却是一个难题。

基于以上的思考以及与业主的沟通，设计师决定把西域元素的某个点进行深度剖析并和现代元素融合应用，以此作为此次设计的重点策略。

土黄色的耐火砖是一种在新疆的建筑出现较多的材质，给人印象深刻。其流行的原因大概是因为耐火砖的颜色与质地比较接近中亚一带传统建筑多用的土坯。耐火砖极少在上海地区使用，因而大量使用耐火砖作为营造西域风情的主要元素，在上海显得十分另类，成为酒吧的重要视觉特征。由于空间变化较少，人们的注意力将会更多地集中在立面的变化上来。从外立面开始，耐火砖就以不同的叠涩方式出现。部分水泥砂浆墙面衬托了凹凸肌理丰富的耐火砖叠涩墙面。而且还使用了许多从老建筑上拆除下来旧地板铺装了大部分地面和墙面，使立面更加丰富、温暖。另外，还在山东淄博定制了一批有浓郁的阿拉伯图案的青花小磁砖，分别运用在了酒吧与咖啡厅的吧台背后的墙上，渲染出了浓郁的西域风情。项目大部分采用了旧家具，并经过重新包布等工艺处理，与部分立面装饰产生了一定的情趣对比。

上述几种主要材质价格都是非常低的，同时也不需要油漆，可以直接安装使用，满足了酒吧边设计、边施工、边营业且无甲醛等污染的要求，最大限度地保证了客人的健康安全。耐火砖表面有很多细毛孔，加上凹凸不平的砌筑方式，对声音起到了很好的吸收和反射效果，前来演出的乐队音乐家表示现场音响效果非常好。还有关键的一点是满足了业主需要"老旧"氛围的特殊要求，因为"老旧"能给人放松的感觉。可以说此设计方案变廉价为宝贵了。

现代音乐有一种被称为"New age"的风格（中文译为"新纪元、新世纪、新年龄"等），是揉合了各国民族风情的混合音乐。该酒吧是以此类音乐的演出为主，出售的饮食也是混合风味的，石榴酒吧就是迎合

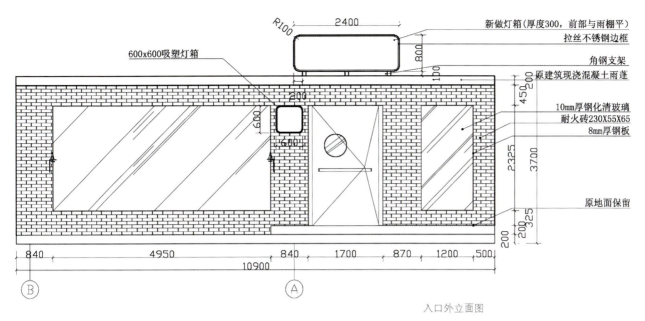

新做灯箱(厚度300,前部与雨棚平)
拉丝不锈钢边框
角钢支架
原建筑现浇混凝土雨蓬
10mm厚钢化清玻璃
耐火砖230X55X65
8mm厚钢板
原地面保留

600x600吸塑灯箱

2400
R100
800
100
200
450
2325
3700
325
200

200
600
600

840　　4950　　840　1700　870　1200　500
10900

B　　　A

入口外立面图

这种经营思路来设计的。

这些，都体现了多元文化混合的时代特征。

This is a project that was designed, constructed and operated in chorus, and changed constantly from the very beginning, as well as its owner partners.

It was said that pomegranates were introduced into mainland China when Zhangqian, an important minister in western Han Dynasty, visited the western part, which now includes Xinjiang Uygur Autonomous Region and some Central Asian areas, such as Afghanistan, and integrates multi-ethnic cultures.

The Pomegranate bar covers 230 square meters and shapes like the number eigh.its smaller outer side faces a streets and on its opposite side is a park, which makes the sightseeing beautiful in the daytime, while the inside is bigger without windows. It provides beverages, western-style foods and performances showed by the original band from daytime to dusk; the majority of its customers are Americans and Europeans who live in Shanghai and Chinese who enjoy the modern entertainments.

The main purpose of investing projects is to gain profit, but how to do it is connected closely with investors' cultural background. For example, investors of Pomegranate bar are Uygur, Hui and Han nationality. "Anar" is the English name of this bar, which is catchy and the pronunciation of pomegranate in Uygur language. So investors hope that it has international manner, and melts with the western style to embody multi-ethnic cultural features and be successful in the market.

The western style is easy to create due to its own western identity, but illustrating what the western area exactly look like is hard for most people, who only know vast deserts, the Gobi steppe, mountains, and Xinjiang folk dances, Thus it is so difficult to reflect the western customs in 230 square meters.

After considering these factors and communicating with the owner, designers realized that the important strategy of this design is grasping one point of western features, studying it deeply and melting with the modern

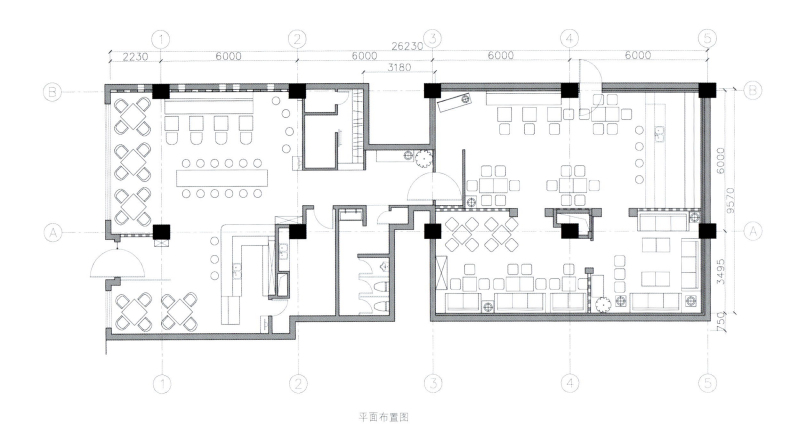

平面布置图

elements.

Khaki firebrick is a construction material, which leaves deep impression and is used frequently in the Xinjiang constructions because of its color and structure that are similar to most buildings in Central Asian made of adobe. However, the khaki firebrick is seldom used in Shanghai, so using lots of firebricks as the key element of western style is very special, which is the essential visual feature of this bar. Due to the limited spatial variation, people pay more attention to its elevation changes. Therefore starting from the facade, firebricks are corbelled on different modalities, and part of the cement mortar walls set off the rich texture bump corbel brick wall. In addition, old floors removed from many old buildings are used to pave the ground floor coverings and wall in the designing process, which enriches and warms the façade. Besides, the designers customize a batch of small blue and white tiles with Arab patterns from

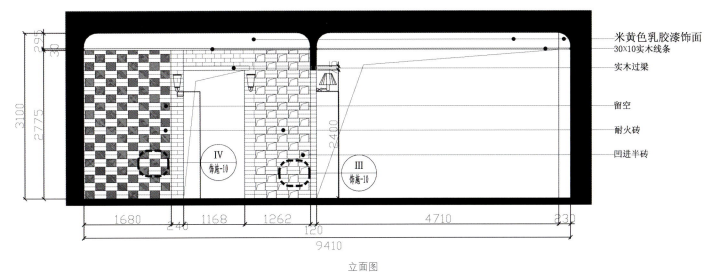

立面图

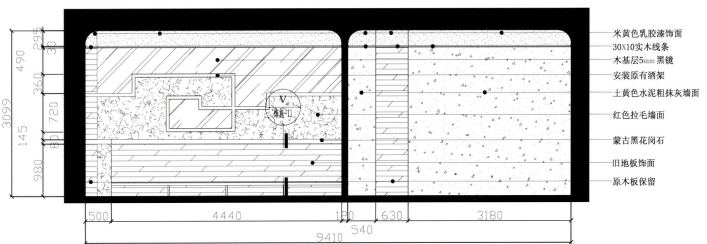

立面图

Shandong Zibo and use them in the wall of the bar and cafe bar, rendering a rich Western style. Most parts of this project use old furniture and process them, such as re-cloth, which becomes a funny contrast with part of the facade decoration.

Prices of the main materials mentioned above are very low, and they also do not need to be painted and can be installed and used directly, all of them meet requirements of designing, constructing and operating in chorus without formaldehyde contamination and ensure the safety of customers to the largest extent. Moreover, the surface of firebricks have many small pores, coupled with the uneven masonry way, making the bar have a very good sound absorption and reflection effects, so the band came to perform said that the live sound is fantastic.

The modern music contains a style,

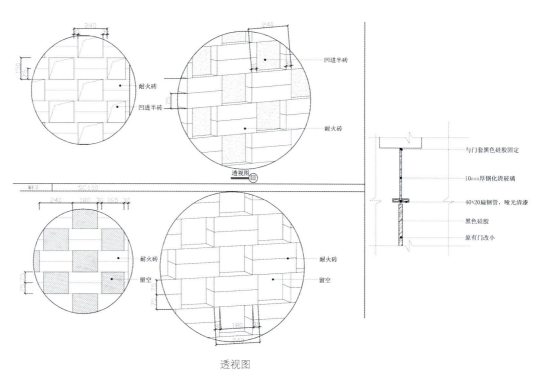

透视图

called New age, which is a mixed music with international style and many different national customs and performed in this bar where provides mixed-flavor food as well. Of course, this design of the Pomegranate bar is in accordance with its business idea.

All of these express the characteristics of this time — multi-cultural integration.

可持续　舒适空间
Sustainable　Comfortable Space | 185

餐饮空间
Restaurant

优胜奖 / Winning Prize

4

项目名称：乐清玛得利餐厅
设 计 者：尹杰
设计单位：杭州意内雅建筑装饰设计有限公司

简约的优雅

酒店的餐饮空间大多金碧辉煌，镶金镀银，无不是想体现一种奢华、高档的空间格调。我们前期设计思考的重中之重，就是要避免雷同，开拓思路，用一种充满新意而又简约的优雅品位来统领全局。

罗马柱、穹顶以及繁复奢华的线条表达不是本案的设计语言。我们尝试实验性地导入装饰艺术风格，混合现代极简的对比色彩和几何化的线条，以及富有情趣的配饰添置，彻底杜绝雍容而又刻意的华丽。独立用餐包厢采用分区化的设计手法，赋予其不同的精神气质与光源搭配，试图让每个人都能远离审美疲劳，最大限度地享受空间艺术所带来的视觉和心理上的优雅、快乐和愉悦。

Simple Elegance

As for hotel dining spaces, resplendent and magnificent seems to be the usual style,and every space strives to embody high end luxuriousness. Not wanting to follow in the footsteps of others, duplicating the already existing, this design strives to truly produce high class simplicity and elegance to encompass the entire design. This was the most important concept for us in the preliminary design stages.

Roman columns, domes and complicated opulent curves are not the most expressive language for this design job, they were abandoned. Experimentally Art deco themes were combined with mixing modern and simple color schemes, geometrical lines and shapes furthermore with rich and interesting accessories were added in order to completely isolate artificial resplendence.

Separate private dining rooms utilise sectorisation giving diverse drive and quality complimented by engineered lighting. This design attempts to reduce the amount of design clutter which in turn lets everyone flowingly and artistically appreciate the elegance and joy of what the design space provides.

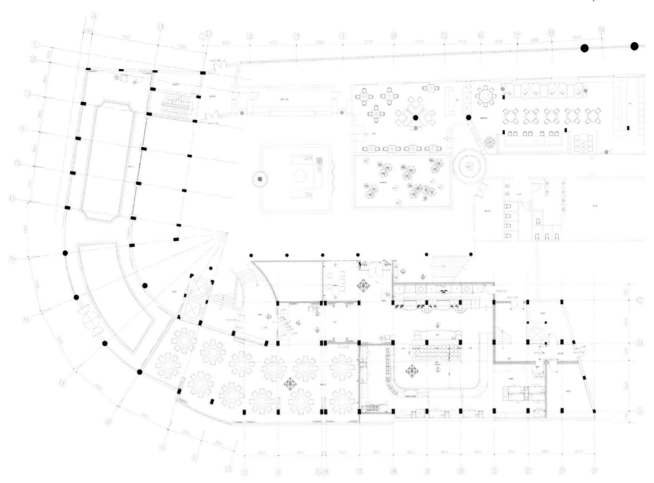

一层平面布置图

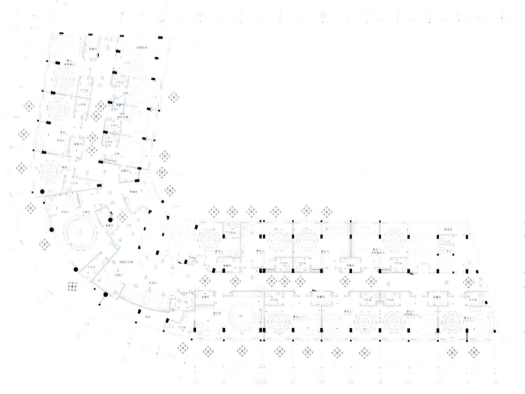

二层平面布置图

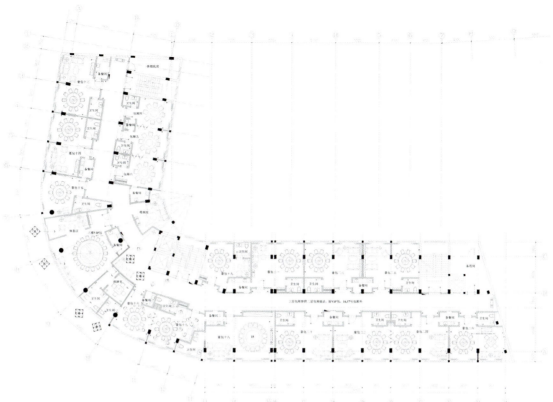

三层平面布置图

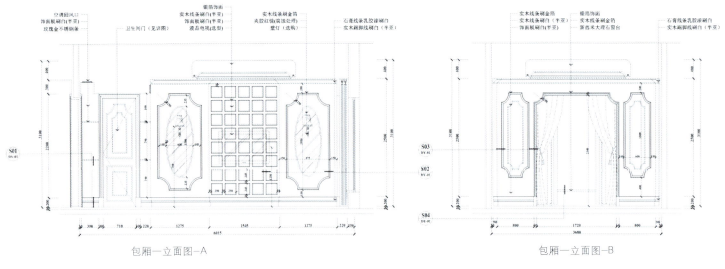

空调回风口
饰面板刷白(半亚)
玫瑰金不锈钢条
卫生间门(见详图)
实木线条刷白(半亚)
饰面板刷白(半亚)
液晶电视(选型)
银箔饰面
实木线条刷金箔
夹胶红镜(高透处理)
壁灯(选购)
石膏线条乳胶漆刷白
实木踢脚线刷白(半亚)

包厢一立面图-A

实木线条刷金箔
实木线条刷白(半亚)
饰面板刷白(半亚)
银箔饰面
新西米大理石窗台
石膏线条乳胶漆刷白
实木踢脚线刷白(半亚)

包厢一立面图-B

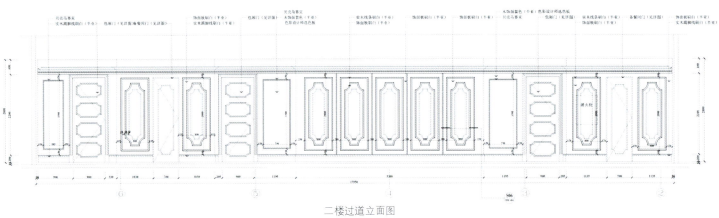

贝壳马赛克
实木踢脚线刷白(半亚)
包厢门(见详图)雕刻门(见详图)
饰面板刷白(半亚)
包厢门(见详图)
贝壳马赛克
木饰面刷白(半亚)
色彩设计师选色板
实木线条刷白(半亚)
饰面板刷白(半亚)
木饰面紫色(半亚)色彩设计师选色板
贝壳马赛克
包厢门(见详图)
实木线条刷白(半亚)
饰面板刷白(半亚)
备餐间门(见详图)
饰面板刷白(半亚)
实木踢脚线刷白(半亚)

二楼过道立面图

产品设计
Product Design

金奖 / Gold Award

项目名称：城市循环
设 计 者：Søren Brøste
设计单位：Furnism A/S

我们是谁

Furnism公司以可持续发展的视野、社会感知度和一种信念来开发，像艺术一样，我们的环境材料可以启发并鼓动我们朝更好的东西去努力。作为设计开发商，我们查明了题目的利益和想法，和艺术家与有远见的设计师一同工作来把理性思维概念融入到独特的商品之中。

Furnism公司把艺术视角和家具及装饰品融合在一起，相信艺术能启发创新思维，并且我们周围的环境应对我们产生积极的影响，让我们创造新的价值。

我们在做什么

Furnism公司提供了价值链上缺乏的环节，在于独特的艺术视角与高端家具和设计物体之间的连接。从创新的理念到可持续性生产和分配，Furnism公司投资并管理着为个性生活创造有意义物品的整个过程。

如何做到

生产的每个物品都依据Furnism公司的指导原则精心选择，以抓住终端产品的艺术视角，满足人类深层次的意思表达需求。

我们并没有使用房间内设形——每个产品打算以一种感性和智慧的方式回应人们，并在用户身上启发个人的诠释。

Edward Burtynsky 提供的图片描述了上海老街区被拆除的过程。其中的木头随即收集堆积起来卖掉，再次运用到我们的城市再利用家具中去。

城市循环实现了利用可持续材料为设计师的主导思想的理念。桌子和椅子是使用上海老建筑拆除之后回收的榆木和其他有机材料生产的。

直到19世纪80年代，大多数上海人住的房子很少有超过两层的。多数情况下，一间房子里住的不只是一家人。常常出现四世同堂的情况，在有些地方，人口密度高达每人仅拥有2～4平方米。

自1990－2002年以来，多达3800平方米的老式建筑和公寓被拆除，为现代化的居住区和商业区留出空间。选用拆除房子的木头是有百年历史的中国榆木，这种木头在中国东部城市的旧式房子中用做室内木板材料和木式结构材料。设计师用这样的木头在现代北欧风情设计中设计了一系列的桌子和椅子。

中国循环文化

和很多西方国家不同，由于环境意识和城市创新的影响，大多数地方的市民都参加于循环利用中。在中国，很多参与者的循环利用的动机仅仅是经济因素。所以，相对来说，几乎没有政府或市政系统用于城市回收。

从地方来讲，中国公民每日会清扫城市大街和村落，收集易拉罐、水瓶甚至老式电冰箱、洗衣机或钢铁、纸质、木质或塑料东西。

循环利用对每个人来说都是有益的。家里没用的废物可卖给收废品的人。我们经常会看到这样的双方在城市某个角落和家门口买卖废品。

Who we are

Furnism develops furniture and accessory objects from a vision of sustainability, social awareness and a belief that, like art, our material surroundings can inspire and motivate us to greater things. As design developers we identify topical interests and ideas, working with artists and visionary designers to develop radical concepts into unique products.

Furnism merges artistic vision with furniture and accessories in the belief that art inspires innovative thought and that our surroundings should infuence us positively and move us to create new value

What we do

Furnism provides the missing link in the value chain between unique artistic vision and the market for cutting edge furniture and design objects. From creative concept to sustainable manufacturing, and distribution- Furnism invests in and manages the complete process of creating meaningful objects for personalized living.

How we do it

Each object we put into production is carefully selected by Furnism's guiding principles, to fulfll a deep-seated human need for meaning by upholding the artistic vision in the end product.

We have no in-house style — each product is meant to resonate with people on an emotional and intellectual level and inspire its own individual interpretation in the user.

Urban Recycling

The pictures by Edward Burtynsky illustrates how the old neighbourhoods in Shanghai are in the process of being demolished. The wood from the housing is then stacked to be sold off and recycled into, among other things, our Urban Recycling furniture.

Urban Recycling realizes a conceptual vision of producing a furniture line for everyday use out of sustainable materials. Its line of tables and chairs are produced using reclaimed elm wood from demolished old-quarter housing in Shanghai as well as other organic materials.

Until the 1980s, most Shanghainese lived in houses rarely higher than two storeys. Often more than one family lived

in a house and a single family might have had as many as four generations living together. In some cases density levels were as high as two-to-four square meters per person.

During the period from1990 to 2002, as much as 38 million square meters of older housing and apartments were torn down to make room for modern residential and commercial properties.

The wood from the demolished housing is century-old Chinese elm. It has been used for interior panelling and for wooden structures in the old quarter houses of eastern Chinese cities.

We have use this wood to design and manufacture a line of tablesand chairs in a contemporary Nordic design.

China's Recycling Culture

Unlike many western nations whose citizens, for the most part, participate in recycling due to a combination of environmental awareness and municipal initiatives, in China the motivation for almost all participants to recycle is purely economic. There are relatively few government or municipal systems for urban collection.

On a local level, Chinese citizens scour the streets of cities and villages daily, collecting anything from soda cans and water bottles to old refrigerators, washing machines or simply anything made of metal, paper, wood or plastic.

Recycling is proftable for everyone. Households with unwanted scrap can sell to collectors. The two parties are often seen haggling deals on street corners or in front of homes.

Prices are calculated by the amount of resource value embedded in the item. Collected items are dropped off at depots. Often located in the suburbs, these depots have, over the years, grown organically to resemble shantytowns. A hardworking scrap collector in Beijing can earn about $125 to $190 US per month.

Our Vision

Our vision for the Urban Recycling line was to design and produce a sustainable furniture Serie constructed from entirely environmental friendly materials.

产品设计
Product Design

银奖 / **Silver Award**

②

项目名称：天然环保竹艺系列
设 计 者：范斌(女士)
设计单位：多维设计事务所

天然环保竹艺灯系列

利用天然环保竹编板材质设计的两款灯具，分别具有现代风格和新东方风格。 现在风格的灯具把"圆形"作为一个基本形，通过"圆"的大小、数量和排列方式达到墙面"浮雕"的视觉效果。 新东方风格：灯具上有纯手工绘制的绵竹年画。但不是直接绘制原图，而是运用"平面构成"的形式。 这两种灯具的尺度都比较大，空间界面强调墙面和落地。灯具已经不仅仅是配饰，而是一个主题的装饰，形成装饰的分量。

Natural ,Environmentally Friendly Bamboo Lamp Series

The two types of lighting fixtures are designed with the natural and environmentally friendly bamboo materials – modern style and new oriental style. The lighting fixture of modern style uses "circle" as the basic shape to create visual effects of "embossment" on the wall by combining circles of different sizes, quantities and arrangement methods. New oriental style: draws the well-know Mianzhu New Year pictures on the lighting fixture. Instead of the directly drawing the original pictures, they are "plan constituted". Both of the lighting fixtures are of big size and the space interface emphasizes on the wall and floor. The lighting fixtures are not only an ornament, but also contribute to the decoration subject with great weighting.

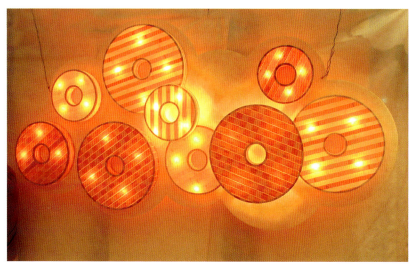

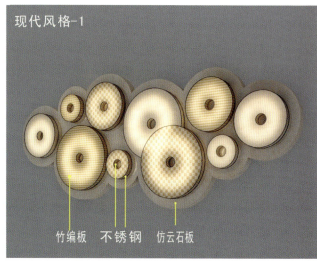

现代风格-1

竹编板　不锈钢　仿云石板

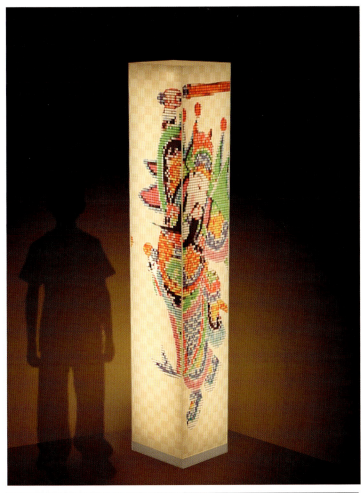

纯手工绘制绵竹年画在灯具上，但是不是直接绘制原图，
而是运用"平面构成"的形式结构传统绵竹年画。

纯手工绘制绵竹年画在灯具上，但是不是直接绘制原图，
而是运用"平面构成"的形式结构传统绵竹年画。

产品设计
Product Design

优胜奖 / Winning Prize

3

"多变"卡接板系列家具

该产品选用零甲醛环保"诺菲博尔板"材质，切割成同一造型的弧形板（弧形创意点来源于小篆体"川"和"水"字最优美的弧度），通过不同的组合方式得到千变万化造型的家具，这种类型的家具既环保，又可以根据空间大小的改变而变化。

"Changeful" Clipping Panel Series Furniture:

The formaldehyde-free "Novofibre" materials are selected and cut into the arc-shaped panels of same moulds (the creative ideal of arc shape comes from the most beautiful radians of "川 (river in Chinese)" and "水 (water in Chinese)" in petit sigillaire). The furniture with various different moulds is achieved through different combination methods. Such kind of furniture is both environmentally friendly and changeable with the space areas.

项目名称："多变"卡接板系列家具
设 计 者：张晓莹、范斌、卢睿
设计单位：多维设计事务所

可持续 舒适空间|199

产品设计
Product Design

优胜奖 / **Winning Prize**

3

项目名称：吉祥如意——"回"
设 计 者：金雨霖
设计单位：上海乐一装饰设计工程有限公司

环保厨房、低碳美学

在传统美学的基础上，设计师从中式宫廷回形元素中获得灵感，以现代的材质和工艺手法对中式宫廷回形图案进行全新演绎，提取其中20×20方形回字格作为最基本的设计元素，适当放大、扭转、变化形态，将传统的艺术气质注入时尚的灵魂之中，效果时尚且雅致，散发出前卫的后现代气息。

具有前卫质感的不锈钢元素和高环保实木的绝妙搭配，把大自然美景与当今潮流趋势都纳入家中，超越了以往选择在装饰效果上的黑白灰风格，5mm金属整体收边设计，增加了设计的时尚感，立体"回形体"弧形倒角，避免了转角的尖锐，体现了整体的美观与现代。再嵌入一些经典白，让整体空间在突显其个性的同时，又让人仿佛置身于无喧嚣、无污染的未来太空厨房中。柜体墙板上回形元素有凸起，有镂空，虚虚实实间让空间隔而不断，相互融合渗透。与柜体相呼应的是空间中心的回立方体，圆润的手法贯穿在门板、台面及储物柜的每个单元，每一处细节也都保持着风格统一，整个空间氛围融为一体。开放式的格局实用大方，现代创新的手法与自然环保的材质完美结合，还原高品位人士的高雅生活，尽显低碳环保生活设计之风范。

面板台面全部采用加拿大枫木并经过 TimberSIL 处理技术。TimberSIL 技术能使木材防火、结构稳定、重量轻、不长真菌、不惧怕白蚁的啃噬，而且不会腐烂。门板采用无拉手设计——嵌入式回形构槽，打造节能专区；智能化的声控式门板，让客户体验非凡生活享受。柜体内外所有的灯具全部采用节能灯管，并通过内置光线感应器，根据客户需求自动控制。

Environmental–Protection Kitchen, Low-Carbon Aesthetics

Based on traditional aesthetics, designers get inspirations from elements of 'Hui' (a shaped of the Chinese character回) in traditional royal court in China which give brand new interpretations to pictures in traditional Chinese royal court by using 20 * 20 square Hui-shape to be taken as the most basic designing elements.Hui–shape By properly magnifying, reversing and transforming, the shape. The effects are both modern and elegant, emitting an artistic sense of fashionable post-modernism.

Surpassing the previous black-white styles, designers use the ingenious combination of fashionable and classy stainless steel elements and high environmental-protection solid wood, trying to integrate the beautiful sceneries and the present trend of fashion in the house. As to the decoration effects, 5mm metal integrated side designing adds a sense of fashion to the design while the three-dimensional 'Hui-shaped' arc-shaped bevelling can help avoid the keenness of the corner that reflects the integrated aesthetic and modern style. The adding of classical whiteness highlights its characteristics that, at the same time, makes one as if position in a future outer-space kitchen free from pollution and noise. The Hui-shaped elements in the wallboard of the cabinet have both embossment and hollowness with dotted and solid spots making the space separated but by no means isolated so that they can be mutually integrated and permeated. Opposite to the cabinet is a Hui-taped cube in the central space that is inserted in each unit of door plank, table and locker with flexibility with each particular maintaining the unified style and the atmosphere of the whole space integrating into one. The open pattern is both pragmatic and liberal with the perfect combination of modern innovative methods and natural environmental-protection materials that can restore the graceful life of people with taste and display the designing style of low-cabon environmental-protection living design.

The surface of the table adopts Canadian maple and are all disposed with TimberSIL technology which makes the woods safe from fire, stable in structure, light in weight, free from fungus, not fearful of termitic gnawing and imputrescible. The door sheet adopts non-handling design namely inserted Hui-shaped slot to forge a special zone for energy conservation; the intelligent voice-control door sheet can make one experience brilliant living enjoyments. All lamps and lanterns both inside and outside the cabinet adopt energy-conservative energy-saving lamps that can be turned on and off automatically in line with requirements of customers through the built-in lighting inductor.

设计方案
Design Proposal

杰出奖 / Excellence Award

①

项目名称：天井——自给自足的建筑
设 计 者：吴楚杰、梁映峰、邓坚
设计单位：A.N.T. Creative Designers Ltd.

天井——自给自足的建筑

居所自古至今都是人类生活最重要的一件事。在由钢筋水泥构筑的当今城市，建筑几乎是千城一面，只能说给我们提供了一个生活的空间，至于住得是否舒适，那就是一个值得思考的问题了。在土地越来越少，建筑成本越来越高，居住成本也随之升高的今天，年轻人寻梦日益艰难。面对日益攀高的楼价和能源消耗，长期的工业文明唤起了人们追求自然、返璞归真的本性，人们越来越要求居住既满足功能需要，又能接近自然。回望中国古建筑，采光、通风、防潮等能源消耗极低，符合当代可持续性需求，因此我们延用了中国古建筑的特点，结合现代科技及材料，设计出这个适合亚热带季风气候城市的自给自足的"天井"。

设计特点
天井

天井可以用来采光、通风、与外界沟通，具有很强的私密性和安全性，可以让人足不出户，便可与大自然中的蓝天白云亲密接触。建筑空间布局以天井为中心，室外的阳光经过天井进入室内，光线变得柔和而温馨；天井内的水池可作为蓄水系统，同时调节室内的湿度和温度，使住宅冬暖夏凉，成为一座"会呼吸的房子"。

结构

整个建筑由30个高2.35米×宽2.35米×深5.9米的标准集装箱组合成一个"井"字结构。每3个集装箱为一户，每户室内面积为41.5平方米，露台面积为21平方米。建筑采用了中国古典家具的接合方式——榫卯结构，这种结构比用螺丝或者铁栓固定更加稳固，具备良好的机动性及可再生、可持续使用性。采用榫卯结构固定便于箱体之间快速组合、拆卸以及维修，使建筑体具有良好的可移动性。

这个"天井"方案，通过设计、组织建筑内外空间的各种物态因素，充分利用自然界的阳光、空气、风和水，使各种自然能源在建筑生态系统内部有序地循环利用，获得一种高效、低耗、少废、少污、生态平衡的居住环境，是符合亚热带季风气候城市的一种生态环保并可持续发展的建筑形式。

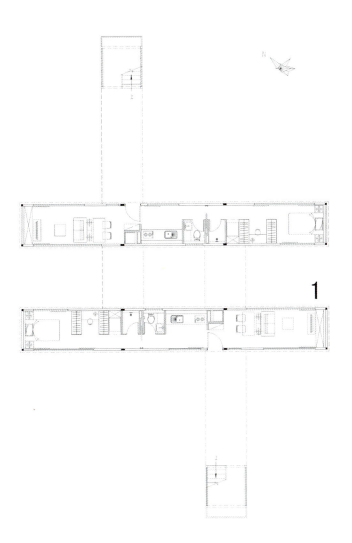

1

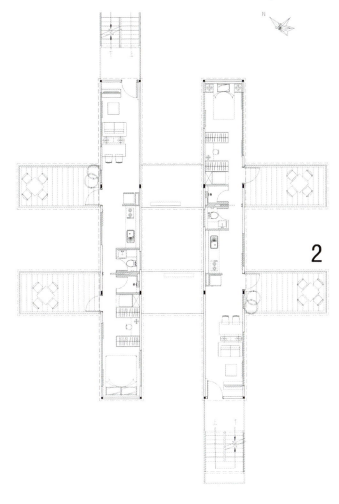

2

Lightwell——Self-Sufficient Construction

Shelter is one of human's basic needs. With the advent of industrialisation, and in today's society, residential buildings have become more a basic living space rather than providing confortable living environment. As cities become denser, land prices and building cost escalating, housing has become less affordable. Couple with higher energy comsumption and continued industrialisation, it becomes apparent that we are now looking for something that is "back-to-basic". Referring tranditional Chinese architecture, it is evident that passive design characteristics such as natural lighting, natural ventilation, low energy consumption are common traits. As such we would like to utilise these traditional design features, together with modern technology and materials, to come up something that suits the sub-tropical urban environment: self-sufficient light well.

Design feature:

Lightwell

Lightwell can be used for natural lighting and ventilation. It is also a mean of connecting the internal with the external while preserving relative privacy and security. In fact, lightwell can serve as a mean of creating an inside-outside feature where built-form speaks with nature.

Sunlight beams though the lightwell gently into the interior. The central courtyard of the lightwell can also serve as an "impluvium", collecting rainwater for recycling usage as well as a naturally occuring air-conditioning system ,cooling down the building in summer and warming up the building in wiater. In fact, we would the building is breathing with the installation of a lightwell.

Structure:

The architecture consist of 30 container boxes measuring 2.35m(H)x2.35m(W)x5.9m(D), arrranged in a hash (#) shape (or the Chinese character '#'). Every 3 containers form a unit; each unit

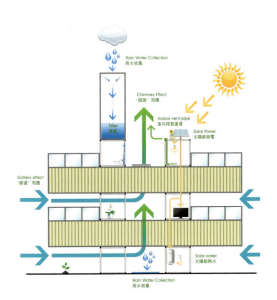

measuring 41.5 sqm internally with outdoor terrace measuring 21sqm. The architecture utilises traditional Chinese furniture technique of mortise and tenon joint connection, which ensures long lasting construction compare to fastener or glue joints Another characteristic of motise and tenon joints is the ability to disessemble and reessemble with speed hence enabling easy maintenance and relocation.

We are utilising design and planning while integrating traditional and modern building technique such as passive design, sustainable building materials and mass produced industrial products to come out with a effective, low energy, low wastage, low pollution and sustainable architecture and living environment that suits our sub- tropical urban context.

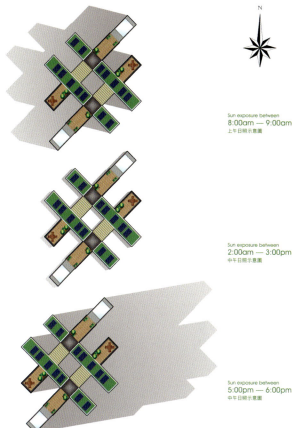

Sun exposure between
8:00am — 9:00am
上午日照示意图

Sun exposure between
2:00am — 3:00pm
中午日照示意图

Sun exposure between
5:00pm — 6:00pm
中午日照示意图

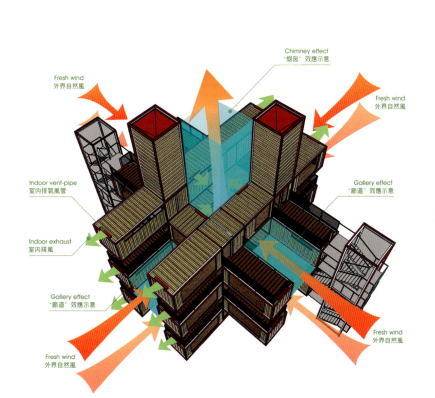

Chimney effect
"烟囱"效應示意

Fresh wind
外界自然風

Fresh wind
外界自然風

Indoor vent-pipe
室内排氣風管

Gallery effect
"廊道"效應示意

Indoor exhaust
室内排風

Gallery effect
"廊道"效應示意

Fresh wind
外界自然風

Fresh wind
外界自然風

Indoor vent-pipe
室内排氣風管

Indoor exhaust
室内排風

Fresh wind
外界自然風

Fresh wind
外界自然風

Calm wind-Hot Pressing Wind
無風時-熱壓通風示意圖

Fresh wind
外界自然風

Indoor exhaust
室内排風

Fresh wind
外界自然風

Indoor exhaust
室内排風

Windy-Hot-Wind Pressing Wind
有風時-風壓通風示意圖

金奖 / Gold Award

项目名称：生态住宅
设 计 者：王瑶
设计单位：厦门理工学院

该住宅选址于厦门的一处竹林山边，可以充分利用本地的资源——黏土以及竹子

建筑

1.基地

这座生态住宅通过分析综合因素，利用地形把建筑基地设在南偏东5°～10°之间的山坡上，建筑形式可以是单体或者联排式的。这样做有利于自然的通风、采光、防晒、排水、躲避暴风雨以及保护耕地等。

2.材料

中国有句古话："石为山之骨，土为山之肉，水为山之血脉，草木为山之皮毛"。将大自然拟人化，讲究"自然与人等价值论"。我们的建筑传承这种尊重自然的思想，就地取材——使用石头、黏土、竹子等作为基本材料（它们主要来自挖掘基地的山坡）。设计师将建筑设计为自然的一部分，使它成为一个有血有肉，可以自由呼吸的"人"。

1）石头：采集的石块可依大小做成基础或石雕装饰物，碎石或者边角料可用做渗透性庭院铺地。

2）黏土：用非烧结黏土砖（No-fired Brick）做主要墙体材料。黏土砖使用红泥等不易耕种的土壤，加入石灰、红糖、糯米以及竹木等进行夯实，以增强其坚固性和防水性。另外黏土可使居住环境冬暖夏凉，亦有呼吸调湿功能，能自动调节室内空气湿度使之相对平衡。在建筑物拆除后，黏土可重新参与到自然循环中，不破坏环境。

3）毛竹：福建省毛竹资源丰富，其生长周期短，只需1～2年就可成材，属于可再生资源。并有一定的强度、弹性和隔热保温性能。经过石灰、熏烤等简单的传统工艺处理便可防潮、防虫。我们将它做成百叶窗、家具以及地板等。

另外根据需要尽量少用混凝土和钢材等材料，因为它们在生产和运输过程中要消耗大量的能源。

3.施工

由于材料大部分是就地取材，所以节省了运输成本。材料的加工使用传统简单的工艺，因此主要雇佣本地的劳动力，有些住户也可以自己完成，从而降低了劳动力的成本，也使施工建设中排放的CO_2大大减少。在施工过程中尽量不破坏周围的生态环境。例如，在采集石头时，注意保护在上面生长的苔藓和地衣；在堆砌石头时使用石灰浆，以便石头可以再次回收利用。

Design Instructions

The location of this residential building is selected near bamboo forest of hillside fields in Xiamen, so that we can take full aduantage of local resources—clay and bamboos.

Construction

1. Location

The location of these ecohouses are chosen at a hillside between 5°–10° south by east through an analysis of countless factors. They can be built alone or together, which is good for natural ventilation, lighting, and drainage, preventing storms and too much sunshine and protecting arable fields, etc.

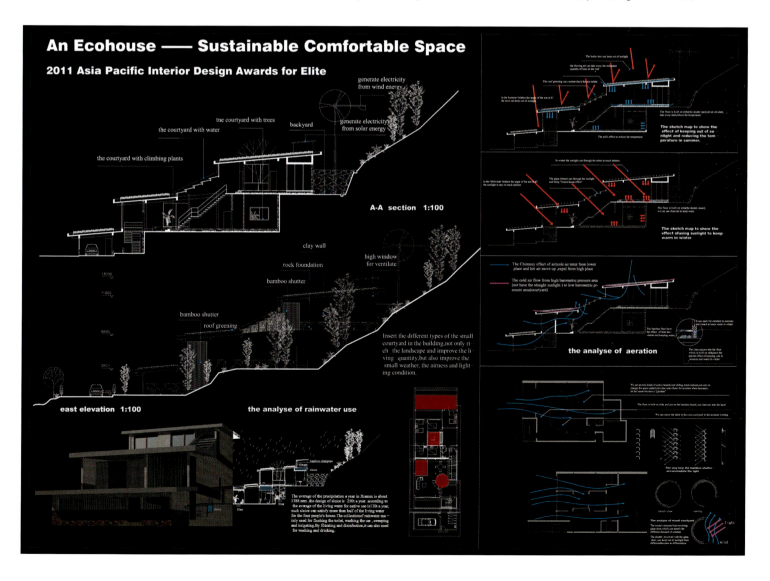

2. Materials

As the Chinese saying goes,"For a mountain, stones are its bones, earth is its flesh, steams are its blood and trees and grass are its skin and hair." In this saying, nature is personified, and people attach great importance to the value of nature, which is equal to that of human. Therefore, we also inherit this idea of respecting nature and draw on local resources——stones, clay and bamboos. We consider the houses as not only part of nature, but a live human being as well.

1) Stones: the stones gathered will be used as bases, made into stone carving ornaments, crushed stones and the leftover bits and pieces can be used for paving the courtyards.

2) Clay: No-fired bricks are used as the main materials to build the walls. These bricks are made from soils unsuitable for cultivation like red earth combined with lime, brown sugar, polished glutinous rice and bamboos, so that they are firmer and waterproof. Moreover, houses made form clays are warm in winter and cool in summer. Thus clays can function as air conditioners and adjust the humility in the air. Even when demolishing, clays can go into nature cycle again, so they are environment friendly.

3) Bamboos: Fujian province is abundant in bamboo resources. Their growth period is only 1-2 years. Bamboos are renewable resources and their characters are superior to those of wood. With certain elasticity and ability to insulate against heat, they are damp proof and worm proof if only treated by a traditional simple technique. We made them into shutters, furniture, floors, drainage pipelines, and so on. Furthermore, only a little concrete and steel are used when necessary in that they consume a great deal of energy in production and transportation.

Construction

Since most materials are local, transportation cost will be saved. Traditional simple techniques are used; so local labor forces, even residents themselves can undertake the construction task, thus reducing cost and emission of CO_2 tremendously. Besides, in the process of construction, the surrounding ecological environment will be protected by all means. For example, those stones having moss and lichen will not be gathered for construction, and lime white will be used in stone laying, so the stones can be reused in the future.

An Ecohouse —— Sustainable C

2011 Asia Pacific Interior Design Awards for El

Design Instructions

The location of this residential building is selected near bamboo forest of hillside fields in Xiamen, so that local resources--clay and bamboos--can be made good use of.

The Climatic and Geographic Conditions

Fujian province in China is located between 115°15' 120°43' east longitude and 23°32' 28°19' north latitude. With subtropical maritime monsoon climate, the yearly average temperature is between 17°C-21°C. The average temperature of the hottest month is 27°C-29°C, while that of the coldest month is between 5°C-13°C. In Fujian, southeast wind prevails in summer and northeast wind in winter. The speed of wind is 3.5m/s on the average. Annual average solar radiation of the whole province reaches 4250/5250MJ/CM², and it enjoys sunshine 55% of the year. It has an average annual rainfall of 1000-2200MM, and March June accounts for 50%-60% of the total. Between July-September,there is much typhoon rain. When it comes to the soil, red earth constitutes the majority. Mountains and hills make up 85% of the total area, so arable land is few and limited.

Aim
1. To reduce the emission of CO2.
2. To create a comfortable, healthy and energy-saving living environment through a most efficient way of drawing on local building materials. In the process of building, using and finally demolishing the houses, we will do our utmost to use least energy and minimize the damage to nature at the same time.
3. To take advantage of natural resources, such as solar energy and wind power.
4. To make good use of topographical advantages, protect cultivated area and surrounding natural environment, so that men, buildings and nature will live together harmoniously.
5. To combine traditional ideas, technology and local resources organically with architecture to create regional ecological buildings. Besides, traditional technology can be handed down.

Means to achieve the aims

1. Low cost: passive low technology and natural materials.
2. Some high technology: solar energy and wind power.
3. Energy saving in construction, little pollution in demolishing and low cost, little emission of pollutants in operation.
4. Respirable artificial pulmonary alveolus: to insert several ecological "courtyards" functioning as small conditioners. Besides the traditional "from outside to the inside" ecological method, a "courtyard" adjusting method that is from the surface to the center is also adopted.

Background Information in Humanities

Chinese people lay stress on geomantic quality. They prefer their houses facing south and beside streams and hills. Chinese five traditional elements--metal, wood, water, fire and earth--can be also applied to architecture, and their harmonious interrelationship is significant.
Moreover, Chinese people pay attention to environmental protection and they have good adaptability to climate variability. For instance, Fujian people like bathing and enjoying the cool in the open air.

Construction

1. Location

The location of these ecohouses are chosen at a hillside between 5° 10'south by east through an analysis of countless factors. They can be built alone or together, which is good for natural ventilation, lighting, and drainage, preventing storms and too much sunshine and protecting arable field , etc.

2. Materials

As the Chinese saying goes," For a mountain, stones are its bones, earth is its flesh, steams are its blood and trees and grass are its skin and hair." In this saying, nature is personified, and people attach great importance to the value of nature, which is equal to that of human. Therefore, we also inherit this idea of respecting nature and draw on local resources-stones, clay and bamboos. We consider the houses as not only part of nature, but a live human being as well.

1. Stones: the stones gathered will be used as bases, made into stone carving ornaments, crushed stones and the leftover bits and pieces can be used for paving the courtyards.
2. Clay: No-fired bricks are used as the main materials to build the walls. These bricks are made from soils unsuitable for cultivation like red earth combined with lime, brown sugar, polished glutinous rice and bamboos, so that they are firmer and waterproof. Moreover, houses made from clays are warm in winter and cool in summer. Thus clays can function as air conditioners and adjust the humility in the air. Even when demolishing, clays can go into nature cycle again, so they are environment friendly.
3. Bamboos: Fujian province is abundant in bamboo resources. Their growth period is only 1-2 years. Bamboos are renewable resources and their characters are superior to those of wood. With certain elasticity and ability to insulate against heat, they are damp proof and worm proof if only treated by a traditional simple technique. We made them into shutters, furniture, floors, drainage pipelines, and so on. Furthermore, only a little concrete and steel are used when necessary in that they consume a great deal of energy in production and transportation.

3.Construction
Since most materials are local, transportation cost will be saved. Traditional simple techniques are used; so local labor forces, even residents themselves can undertake the construction task, thus reducing cost and emission of CO2 tremendously. Besides, in the process of construction, the surrounding ecological environment will be protected by all means. For example, those stones having moss and lichen will not be gathered for construction, and lime white will be used in stone laying, so the stones can be reused in the future.

Plants
Considering their adaptability and low cost, local plants will be selected to be grown in the courtyards and on the roof. Broadleaf deciduous trees can be shelters from sunshine in summer, and they will let the sunshine into the room in winter. And some plants, which can absorb CO2 and SO2, can be grown. (like Albizia julibtissin, Trachycarpus fortunei, Rosa chinehsis ,Hibiscus rosa-sinensis) For the residents convenience, vegetables, herbs and fruits can be cultivated too. Moreover, ornamental plants with different flower seasons can be selected. Plants are not only good for the small climate and air, but also provide a pleasant environment for insects and birds.

1. house
2. path
3. river
4. sidehill

site plan   1:1000

fortable Space

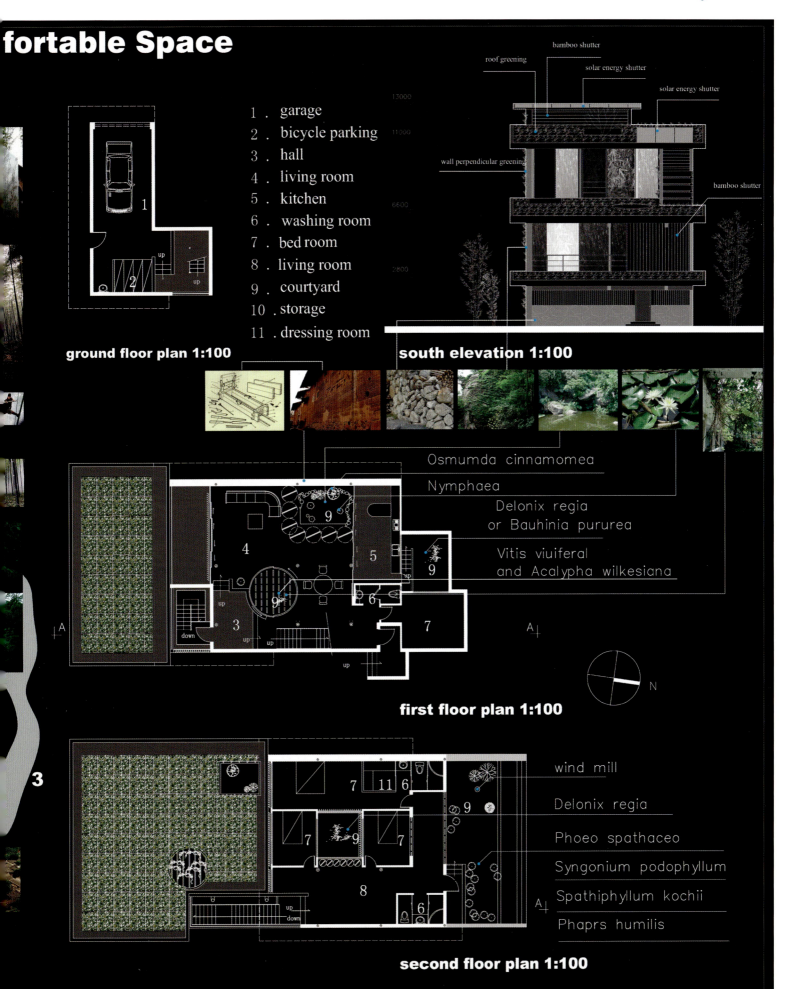

1 . garage
2 . bicycle parking
3 . hall
4 . living room
5 . kitchen
6 . washing room
7 . bed room
8 . living room
9 . courtyard
10 . storage
11 . dressing room

ground floor plan 1:100

south elevation 1:100

roof greening
bamboo shutter
solar energy shutter
solar energy shutter
wall perpendicular greening
bamboo shutter

13000
11000
6500
2800

Osmumda cinnamomea
Nymphaea
Delonix regia
or Bauhinia pururea
Vitis viuiferal
and Acalypha wilkesiana

first floor plan 1:100

N

wind mill
Delonix regia
Phoeo spathaceo
Syngonium podophyllum
Spathiphyllum kochii
Phaprs humilis

second floor plan 1:100

设计方案
Design Proposal

银奖 / Silver Award

3

项目名称：唐风温泉主题度假酒店
设 计 者：史新华
设计单位：浙江建工建筑设计院有限公司、易境环境艺
　　　　　术设计所

武义唐风温泉主题度假酒店

武义唐风温泉主题度假酒店因拥有天然温泉资源而闻名。为了打造高端主题度假酒店，设计师特别从文化高度进行整体策划并通过室内设计来体现。设计以唐开元年间李隆基和杨玉环的风雅故事为文化基础，配以宫廷贵族和文人雅士所崇尚的文化场景元素，再结合现代高端主题度假酒店的经营理念，形成一个〝根植于传统文化，迎合现代休闲理念〞的风格定位。

贵宾楼接待大堂的设计意象来源于对中国传统建筑〝堂〞的理解。在有限的空间条件下，充分运用了设计语言比例关系和玄妙雅致的灯光氛围，再现一种〝唐风弥漫、古韵缭绕、高阔优雅〞的厅堂氛围，给贵宾的情感上带来了高度的归属感和期望感！

全日餐厅的设计之所以结合了自助餐厅和主题茶吧的混合氛围，是因为餐厅窗外拥有最佳的户外温泉主题景观，这使得这一空间需要内外呼应的两种高品质资源，因此在室内设计上以舒展大气为主。

客房设计以楼层为单位，分别以仕女文化、唐诗文化等不同主题使客人拥有无限的精神体验。室内空间甚至融入唐人品茶、对弈的场景，令人心旷神怡，梦回盛唐。

整个项目的设计特别注重选材和技术上的环保和非工业化，强调和整个温泉山庄地理环境相协调。例如，采用了当地的砂岩（自然面）做主要的装饰材料；木饰面工艺采用传统古建筑油漆工艺，与整个室内外建筑的古典风格相吻合；照明设计则极力淡化城市豪华酒店的理念，极少使用大功率射灯光源，更多结合自然光理念，采用ＬＥＤ光源结合反光槽技术，营造符合传统文化含蓄、质朴、典雅的艺术氛围。

Design Specifications of Wuyi Tanfeng Hot Spring Resort Hotel

Wuyi Tanfeng Hot Spring Resort Hotel is famous for its natural hot springs. In order to build the concept of high-end theme resort hotel, we make overall plannings especially from the height of culture, and reflect it integratedly through interior design. The design concept finds cultural origins from Changan Lishan Huaqing Pool of Tang Kaiyuan, and then extracts the elegant stories about Li Longji and Yang Yuhuan around Tang Kaiyuan, using cultural scene elements advocated by court nobles and literati, combined with the business philosophy of modern high-end theme resort hotel, to form the style positioning at "rooted in traditional culture and meeting the modern concept of leisure".

The design image of grand reception hall comes from the understanding of traditional Chinese architecture "tong". Under the conditions of the limited spatial scale, we make full use of the proportional relationship of the design language and mysterious and elegant lighting atmosphere, to reproduce a hall atmosphere of "diffusing Tanfeng, ancient rhyme-filled, high, wide and elegant", giving honored guests the sense of belonging and expectations from the height of the emotion.

The design of all day restaurant combines a mixture atmosphere of cafeteria and theme tea buffet. It is because the platform out the window of the restaurant has the best landscape of outdoor hot springs, that makes this space have high-quality internal and external resources, and thus the interior design also

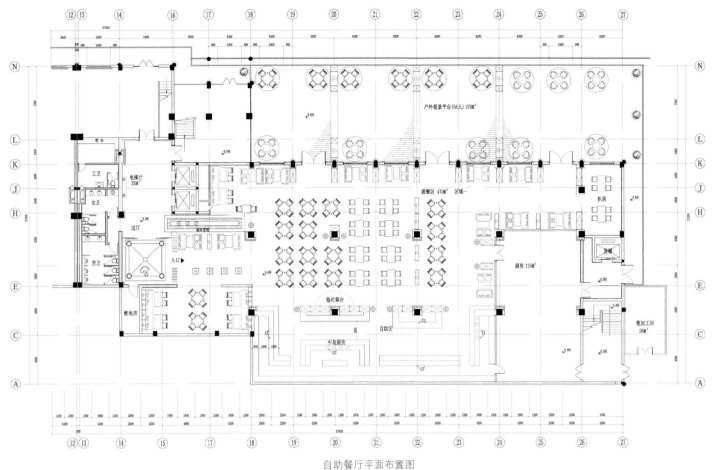

自助餐厅平面布置图

characterizes with stretched and generous.

Room design gives floors different themes(like Ladies culture, Tang poetry culture, etc), to make guests have boundless spiritual experiences; interior space even integrates with the scenes of Chinese tea and chess, which is refreshing and makes people dream back to Tang.

The design of the entire project focuses on environmental protection and non- industrialization in the selection of materials and in tachnology, emphasizing the necessity of coordination with the physical environment of the whole Hot Spring Villa.Such as using local sandstone (natural surface) to be the main decorative materials; wood veneer craft takes traditional painting, fitting with the classical style of the whole building; lighting design strongly dilutes the concept of urban luxury hotel, and rarely uses high-power spotlights source, instead, combined with the concept of natural light, uses LED light with the technology of reflective groove to create a subtle, simple and elegant art atmosphere in line with the traditional culture.

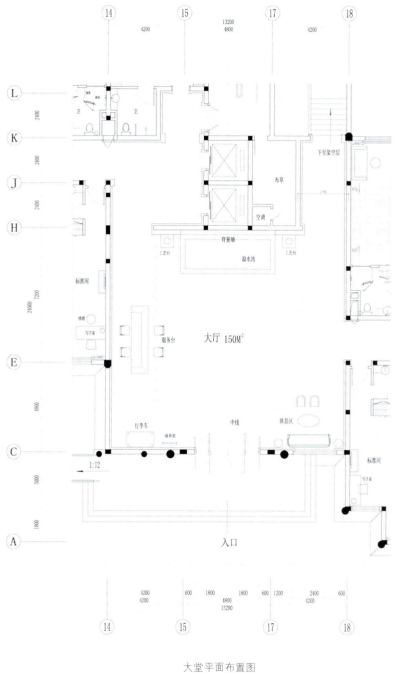

大堂平面布置图

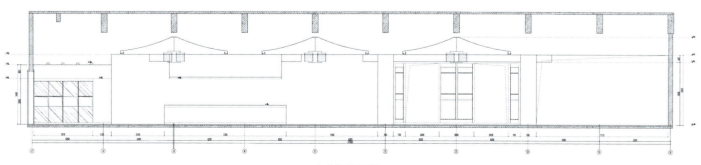

自助餐厅立面图

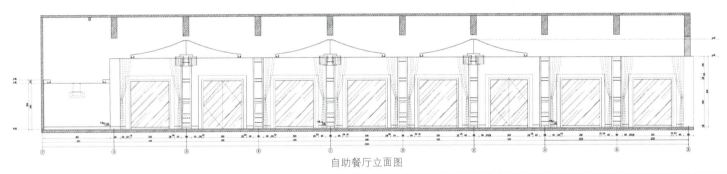

自助餐厅立面图

标准间卫生间

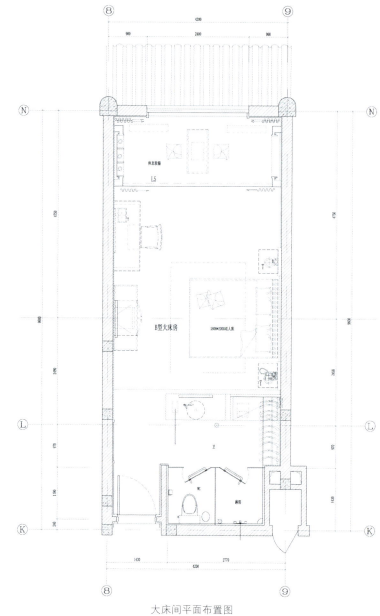

大床间平面布置图

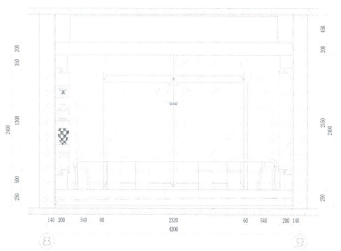

大床间立面图

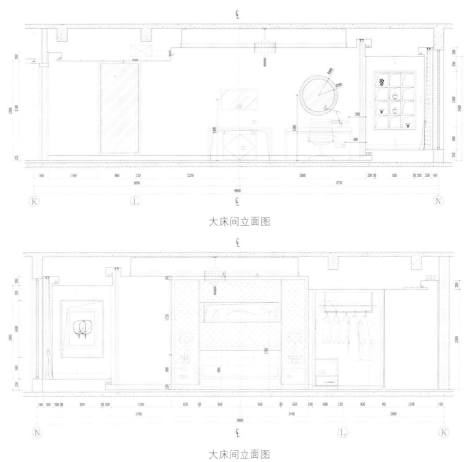

大床间立面图

大床间立面图

沁芥兰中式时尚连锁快餐店

这是一家连锁快餐品牌店面，空间主要以健康、绿色环保为主题。在设计方案中，除了以绿色调为主要视觉表达以外，还选用了零甲醛低碳和可降解的材料，且部分面材以安装方式来减低现场的二次污染。

主题装置以"饭勺"为基础元素，突出了空间主题又富有情趣。

LED灯的使用使空间更节能；厨房打开窗的玻璃设计使消费者了解到食物的制作流程。这是一家社区性的家庭餐厅，且空间的互动性较高；VI设计根据已完成的室内视觉元素来制作其标志。

项目名称：沁芥兰中式时尚连锁快餐店
设 计 者：张晓莹、祝鹏
设计单位：多维设计事务所
材　　料：防火板、玻璃、仿古地砖、乳胶漆
面　　积：400 平方米

Qin Jielan Chinese Fast Food Chain

It is a fast food chain which focuses on healthy and friendly environment. In this schematic design, not only that green is used as the main color, but the adopted materials are formaldehyde-free low-carbon and degradable and partial surface materials are installed in a way to reduce the secondary pollution in site.

"Spoon" is the basic element in the subjective decoration art, not only highlights the space subject but also outives the space atmosphere.

LED lamps are efficient in energy saving and windows in the kitchen are open which provide customers with an access to know the flows in kitchen. As a community restaurant focuses on families with quite high space interactivity, VI design makes its logo conversely according to the interior elements after some interior visual elements are completed.

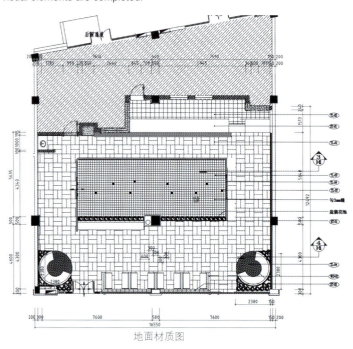

地面材质图

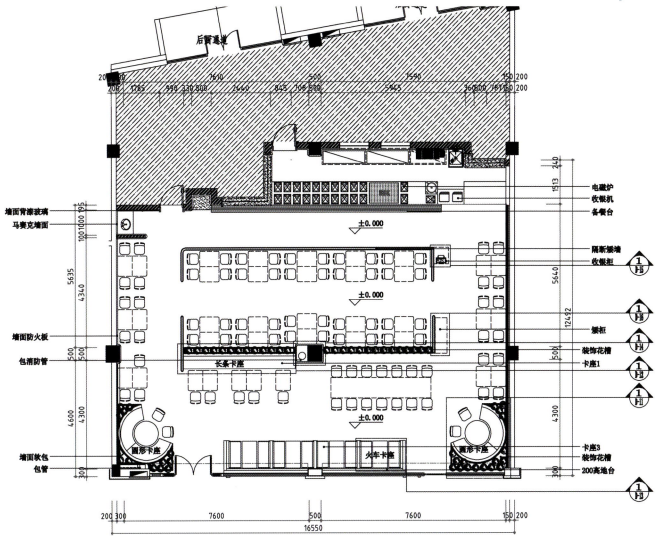

墙面背漆玻璃
马赛克墙面

墙面防火板

包消防管

墙面软包
包管

后厨通道

±0.000

±0.000

长条卡座

圆形卡座

火车卡座

±0.000

电磁炉
收银机
备餐台

隔断矮墙
收银柜

矮柜
装饰花槽
卡座1

圆形卡座

卡座3
装饰花槽
200高地台

平面布置图

主题装置艺术以"饭勺"为基础元素，
突出了空间主题又富有情趣

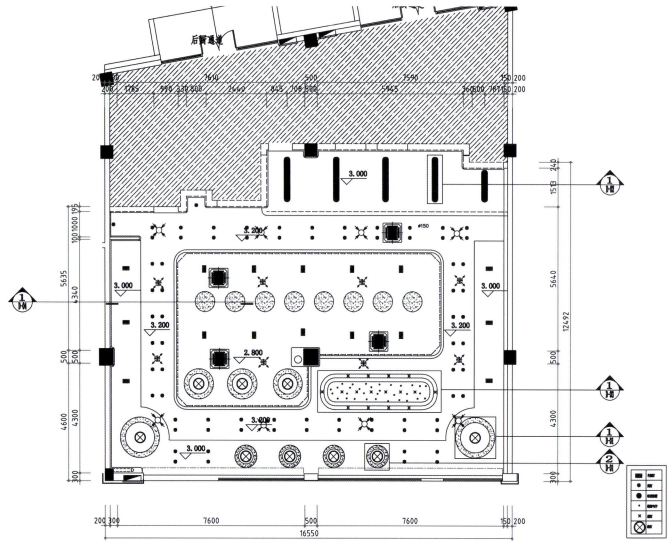

天花布置图

设计方案
Design Proposal

银奖 / **Silver Award**

3

项目名称：四川·成都佳士科技综合办公楼室内装饰
项目面积：7300 平方米
设 计 师：王玉珏

佳士科技综合办公楼

装饰设计思路：根据对成都佳士科技有限公司的第一印象以及之后对于贵企业的更加深入的了解，我们对设计使用的颜色有了初步定位：黑、白、灰、橙和金属色（钢铁）。

其中黑象征厚沃、坚实的土地，暗指我们脚踏实地的企业态度；白象征干练、洁净的白云，飘于空中与坚实的大地遥相呼应；灰象征严肃、严谨，表明我们做事情的态度；橙则是我们企业的代表颜色，热情、希望、丰收、喜悦；金属色象征钢铁的意志，并能够让人予以信赖以及依靠。这几个颜色的组合体现出企业的宗旨："专业、专长、专注"，并且将此宗旨贯穿到我们整个办公场所的每一个空间。

方案特点：在简洁、大方的整体环境当中，着重运用钢铁的本色以及独特的焊接拼接技术，显示出企业主营钢材焊接项目，传递出企业的宗旨"专业、专长、专注"，并使空间具有独一性。

整体感强烈的空间当中运用企业的代表颜色——橙色进行点缀，让整个空间都充满灵通感以及跳跃感。

人性化与空间合理化、人体工程学完美结合。空间除了当中造型、颜色的运用等亮点以外，在空间隔断、造型材质的选用上也是别具匠心的。顶面模仿天光的造型、墙面石膏板与玻璃隔断的完美结合、连廊过道与外部环境的呼应……无不体现出我们企业的人性化、环保性、环境的不孤立性，也是我们方案的最大亮点。

设计构思：将我们的方案主线引申到我们的每一个空间当中，让每个空间得到完美的融合。我们的目的就是：使我们的建筑与整体佳士科技工业园区相结合，与整体的外立面设计相融合，与整个佳士科技企业相呼应。即使对企业一无所知，踏入我们的空间就能被说服和感染。

视线设计：在办公区域当中，我们尽可能地考虑侧光源，在有效地避免了太阳光对肉眼的直射和逆光侵扰的同时，也考虑到了员工工作时所产生的反光以及视觉疲劳。对于过道我们考虑采用自然光源。

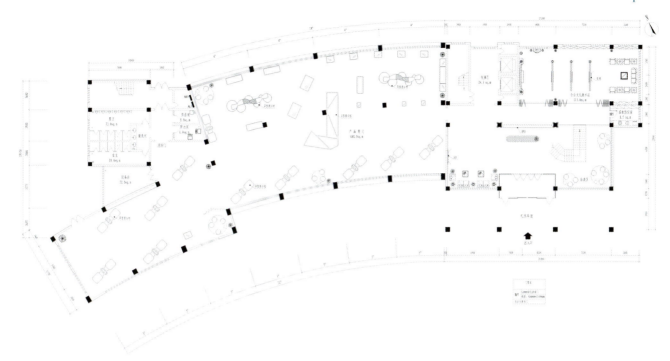

一层平面布置图

Jasic Technological Office Building Design Report

Decoration Design Idea: according to the first impression and future understanding of Chengdu Jasic Technology Ltd., we decide Black, white, gray, orange and metallic color (steel) to be the color orientation.

Black, thick and solid ground, represents the surefooted company attitude. White, pure and simple as the white cloud floating, echoes the solid ground in the air. Gray, serious and precise, indicates our working manner. Orange, the deputy color of our cooperation, symbolizes passion, hope, harvest and joy. The metallic color

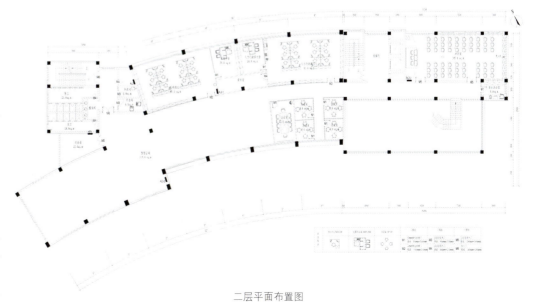

二层平面布置图

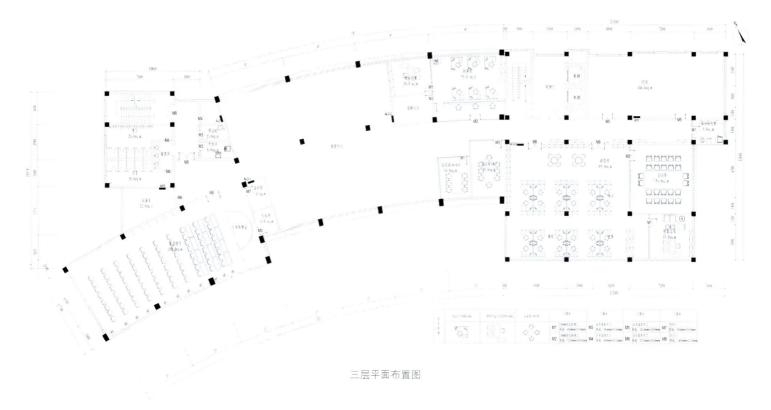

三层平面布置图

suggests the steely will, steady as steel, firm as metal, and solid as stone. It's trustable and reliable, just as our cooperation aim: professional, expert and devoted. We impenetrate the aim into the office space by using the steel characteristic and its unique splices model to present the peculiarity and uniqueness of space.

Project Speciality: to show the peculiarity and uniqueness of space, we will focus on using the original color of steel and by special jointing and splicing technique to handle the steel and armor plate which will be as delicate as a perfect art piece, and therefore show the professional, expert and devoted standard.

Orange, which is the major color to present the sense of whole, will be used to embellish the space, emphasize the atmosphere, and decorate the space more actively and vividly.

The humanity will be perfectly combined with space rationalization and human engineering. Besides

special sculpture and color management, space separation and material variety are also our highlight points. The top is designed as daylight sculpt, and wall is combined perfectly with plasterboard and glass obstruction. And the aisle echoes with the outside environment. All of the design is highlighted and based on the cooperation's humanity nature, environmental protection, not to isolate the environment.

Design Conception: we will extend the case major concept into every space, to insist overall interlude and impenetration. All the space will interact ideally. Our purpose is to echo with the overall Jasic Tech Zone and façade of the design, and of course the Jasic company. It will provide a suitable and attractive environment to all the people stepping into the office.

Line of Sight design: in the working area, we make more effort on the side lights. It can not only prevent direct sunshine and back lights' intrusion but also help staff to reduce glisten harm and relax eyes. As to the aisle, we will make full use of the nature light.

四层平面布置图

设计方案
Design Proposal

优胜奖 / **Winning Prize**

4

项目名称：北京晋城会所
设计公司：香港联合国际设计有限公司
设 计 师：赵洪亮

北京晋城会所
可持续设计理念

可持续设计，又称绿色设计，是一种以符合经济、社会及生态学三者融合为方针的设计方法。可持续设计的范围跨度极大，小至各种日常生活用品，大至建筑设计、都市设计乃至整个地球的自然环境。

简单地说，一个设计在投入使用后，不会对环境造成危害，在使用周期结束还能够回归自然，甚至回馈大自然，同时也能保证产品质量，就叫做可持续设计。如今，能源短缺、污染严重等问题一直摆在人类面前，如果要使这些问题有所减缓，人们需要学会在生活实践中更为合理地协调与自然的关系。

这就是可持续设计的意义所在。

在文明累进的过程中，人类对自身与自然关系的理解也在持续变化，从而影响到人类营造生活环境的方式。西方的启蒙运动在推动科学与哲学突飞猛进的同时，也向人们灌输了人类拥有管理自然的权力，可以主宰自然环境的思想。自然界被看做是一个服务于人类的大机器，这一将人与自然割裂的思想成为了近代环境恶化的主因之一。

但在长期的实践中，人类对生命和自然的认识逐渐有了本质性的进展，发现自然系统之间的各个元素并非独立，而是彼此环环相扣，人类仅是该复杂系统中的一个环节而已，我理解室内设计也是自然系统中的一个元素。人类生存依赖的是整个地球生态系统的健全。

秉承这样新的设计观，本案的可持续的概念慢慢浮现。在营造舒适空间的同时，要既符合当代人的需求又不损害后代的利益。在此思潮推动下，本次北京晋城会所空间设计概念也应运而生。

环保创意、低碳创意

北京晋城会所设计从诞生之日起，就包括了"环境"、"人"、"材料运用"等环节，着重强调人与环境的关系。在产品的整个生命周期内，要考虑产品的环境属性（可拆卸性、可回收性、可维护性、可重复利用性等），并将其作为设计目标；在满足环境目标要求的同时，保证产品应有的功能、舒适度、使用寿命和质量。

北京晋城会所，白天会所从不开灯，最大程度利用自然光线，从而减少三分之二的二氧化碳排放；会所内所有的笔都采用生物可降解材料制作，纸则使用可再生纸浆；会所木作选用低排放型开放漆；会所石材选用石材切面工艺直接安装，不做表面抛光和结晶工艺处理；甚至连卫生间内都加强了马桶冲水的力量，以节约45%的用水。

在可持续设计导向下的建材产品有可能成为世界的主导产品，而可持续设计的标准也将成为生产行为的规范。

在未来，可持续意味着人类对永恒、持续发展的期望，所有的概念、方法、技术都是为了达成这个期望的

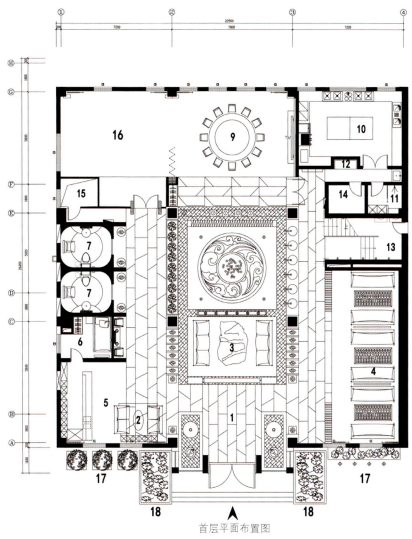

首层平面布置图

手段。而最根本的，还有在于人类本身的思考、欲望和最后的价值取舍。

本案是建筑师赵洪亮的原创设计作品，功能布局科学合理、交通流线设计流畅，整体空间运用可持续设计手法融入中国晋商设计元素，设计手法和设计语言也不仅要具有原创性、唯一性、环保性、可持续性，并有信息化、视觉化、现代化与时尚性和艺术性并存。

Beijing review made by the

Sustainable Design Concept

Sustainable design, also called green design, is a kind of method that throngh a combination of the economic, social and ecological scholars. Sustainable design range span is great, small to all kinds of articles for daily, from the large building design, urban design and the whole earth's natural environment.

Simply put, a design, will not put into application in environmental harm, in use cycle still can return, and even end feedback nature, also can ensure the quality of products, is called a sustainable design. Today, energy shortage, serious pollution have been an issue in which human before, if want to make these problems, people need to learn to slow down in life practice more reasonable coordination and natural relations.

This is the meaning of sustainable design.

Civilized progressive, in the process of its own natural human understanding of relationships

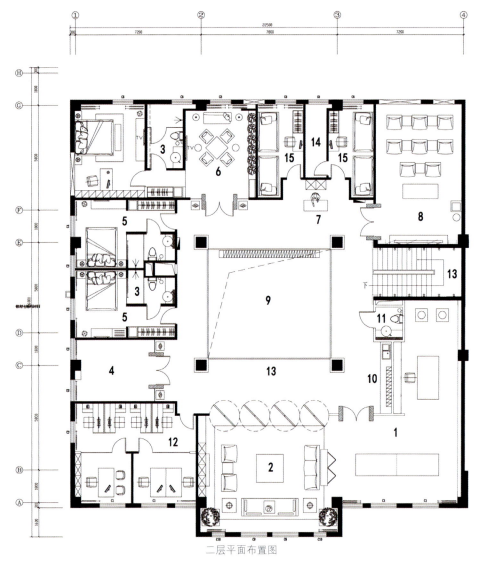

二层平面布置图

continues to change, thus affecting human build the living environment of the way. Western enlightenment movement in promoting science and philosophy by leaps and bounds and, at the same time, to engraft to people with the human nature management power, can independence natural environment of the civil thoughts. Nature is seen as a service to the human big machine, this will be a man and nature of the separate thought became one of the leading modern environmental degradation.

But in the long-term practice of human life and nature, the understanding of the essence is beginning to progress, found that between each element of natural system is not independent, but each other linked together, humans only the complex system in a link just, I understand the interior design is also an element of natural system. Human survival depends on the whole earth ecosystem sound.

Take this new design concept, the case of the concept of sustainable slowly emerge. In space at the same time, create a comfortable with that contemporary and does not harm the offspring of demand in the ideological trend. Interest ", pushed by the Beijing review made by club space design concept also arises at the historic moment.

Environmental protection of creativity, low carbon originality

Beijing review made by the design from birth day, including the "environment", "human", "material use links such as" emphasized the relationship between human beings and the environment. In the whole life cycle of products, product environmental attributes (considering the disassembling, recycling sex, maintainability, reusable sex etc) as their design goals; To meet the requirements of the target in the environment at the same time, to ensure that the products its function, comfortable, service life, quality.

Beijing review made by the design of the club during the day, the light, the greatest degree never use of natural light, so as to reduce carbon dioxide emissions; two-thirds of the Club all the pen USES biodegradable material is made, the paper USES renewable pulp; The wood to make choose low emissions open type paint; The stone cut stone material selection process installed directly, don't do surface polishing and crystallization process; Even in the toilet is to strengthen the toilet flush water power to save 45% of the water.

Under the guidance of sustainable design in building materials product might become the world's leading products, and sustainable design standard also will become the standard production behavior.

In the future, means of eternal human sustainable, and continuous development of expectations, all of the concept, methods, technology are expected to reach this means. And most of all, and lies in human itself thinking, desire and the value of the final choice.

This case is the architect Zhao Hong Liang original design of the works, the function layout of scientific, rational, traffic streamline design fluent, whole space sustainable design technique used in the Chinese traders and design elements, design methods and design language also should not only are original, uniqueness, environmental protection, sustainable, and information, visualizations, and modernization, and fashion and artistic coexist.

优胜奖 / **Winning Prize**

4

项目名称: 创意生活
设 计 者: 漆 立
设计单位: 湖南金煌建筑装饰集团湘潭分公司

临水而居 沐浴阳光 创意生活

设计风格简洁、纯粹、流畅, 集中体现了现代主义"少即是多"的思想, 而色彩以黑白灰为主, 张扬, 年轻。

本案为创意户型, 从形式上彻底改变了传统生活空间的概念。空间结构让居住者对生活有一种全新体验, 同时把水和自然景观引入家中, 让居者感受到"临水而居, 沐浴阳光, 创意生活"的惬意。

注重保留空间的特点: 横向全落地窗, 空间充分与大自然有机结合, 让休闲、创作都感受着大自然的纯真和伟大。室内设计采用了极简现代手法塑造主体风格, 同时结合LOFT特征以呼应创意人士的心理诉求。洁白的自浆流平砂地面如水面般映射出白的墙、黑的墙、窗外的风景。充足的自然光极大地减少了照明用电, 而光面的百叶帘有效地调节了白天的室内照明。低碳环保创意与实用性都很强。

客厅、卧室交界处的电视墙是悬挂式的, 成为空间中的艺术装置, 也成为空间的主体。电视墙也可一物几用, 背面是书架, 使客厅与卧室分离, 既有视觉冲击力, 又宜于家居生活需要。钢锁链从天花板顺着地吸引力向下层延伸, 贯穿了整个空间, 又强调了空间的LOFT特征和力量感。简炼纯粹的空间配以钢锁链、大理石, 注入了创意和力度。

餐桌、橱柜、操作台、洗簌台浑然一体, 犹如空间中的一条纽带, 连接了主卧与客厅以及厨房。主卧的吊床考虑了户外景观, 拥有面水而居的开阔视野。主卫是开放的, 全无分隔, 把浴缸放在观景的主卧阳台上, 体现了创意人士的求异思维。

深蓝色和灰色的墙、白色的顶、白色的地面, 结合黑色和白色的钢锁链, 形成空间的黑白灰语汇, 用极简工业主义体现了独特的审美, 诠释了创意的生活。

The Seaside Apartment—Creative Life with Sunshine

The design style of this case is concise, pure, and smooth, concentrating on the "less is more" modern socialist thought, with black, white and grey as the main color, presenting youthfulness and conspicuousness.

As a creative-sized apartment, this case changed the traditional life space concept completely via the form. Space structures give birth to a new experience of life to habitants. At the same time, seaand natural scenery are brought into the house, giving the person a feeling of creative life that they are stepping into the sunshine and sitting along the sea.

In order to keep the space characteristic, the transverse space is adequately with French window, making the full combination of space and nature. Meanwhile, the great and innocent nature can be found in leisure and creative works. Interior design uses an extremely brief modern means to form the main style, simultaneously; it combined with LOFT style features to echo the creative personage psychological demand. White self-leveling ground reflects the white and black wall, and the scenery outside the window as water surface. Enough natural light greatly reduced the lighting electricity, Venetian blinds toward the

sunny side effectively adjust indoor lighting during the day, and thus it has low carbon and practical applicability.

The TV wall in the intersection of living room and bedroom is suspension type, contributing to the space installation art, becoming the space subject. TV wall can also have other usages: with the bookshelf at back, it separated the living room and bedroom; both created a visual impact and household life need. Steel chains extend down from the ceiling to the lower

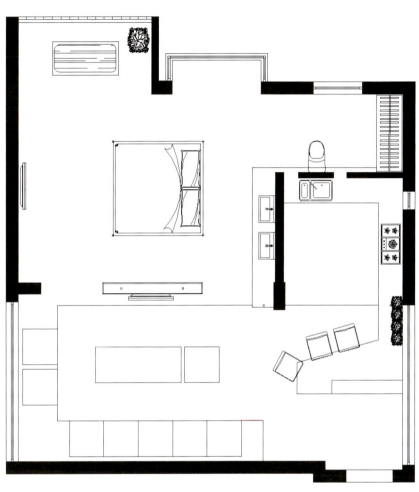

平面布置图

levels to across the whole space by the gravitation, which emphasized the space LOFT characteristics and the sense of power. Simple pure space, matched with steel chains, marble, injected originality and strength.

Dinning-table, cupboard, operating floor, washstand all blended into one harmonious whole, just like a belt which connected the master bedroom and living room as well as the kitchen. The hammock in the master bedroom blends in the outdoor landscape, owning the wide vision of living along the water. The main bathroom is open-type, with nothing to separate it, putting the bathtub on the balcony of the main bedroom, reflected the divergent thought of creative person.

The dark blue and grey wall, as well as the white roof and floor, combined with dark & white steel chains, formed achromatic colors of the space. It expresses unique appreciation of beauty and annotates the creative life with extremely brief industrialism.

设计方案
Design Proposal

项目名称：北京鸟巢艺术酒店
设 计 者：Enrico Taranta
设计单位：上海半舍建筑装饰设计有限公司

北京鸟巢艺术酒店

随着北京鸟巢体育馆的竣工，许多设计爱好者纷纷期盼这栋独特建筑物的下一个动作。既然近期的体育目标已达成，那么远期的经营目标则想当然成为接下来的首要考量。北京鸟巢艺术酒店项目提案，是结合了独特的设计理念，并落实远期经营策略之最佳代表。

根据我们的提案，北京鸟巢艺术酒店的设计灵感来源于鸡蛋的形状。从超现实主义艺术家，尤其是埃舍尔的灵感当中，我们将不可能实现的艺术主题转移到建筑层面。由于使用了最新的设计和参数化软件，我们开发了从克莱因瓶表面一个特定等式的不可能表面。结合"窝巢"的概念与数学算法，酒店大堂就这样诞生了。大厅里有一个弯曲状的楼梯，并且有一些鸡蛋形状的椅子。颜色的色调非常简单干净，是纯白色并且加一些紫色，楼梯的形状也非常独特，人们可以将这里当做举办时尚表演的空间。北京鸟巢艺术酒店的饭店的主题也围绕鸡蛋这个形状，这里有许多弯曲形状的竹藤家具，并且前方的吧台也是圆形的。酒店里还有一个日式酒店和一个温泉浴室，也是以鸡蛋形状为主题的。公寓套房也设计得非常优雅，起居室里有一道转墙，并且这道墙上装有LED。洗手台是开放式的，位于房间中而不是卫浴内，卫生间门外墙上的绿色垂坠的树叶也是一个重要的设计特色。

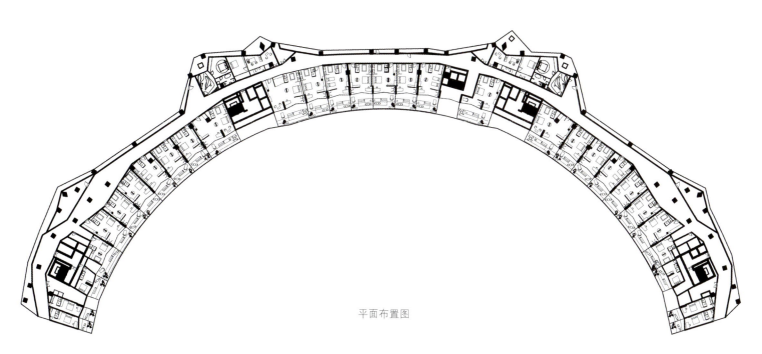

平面布置图

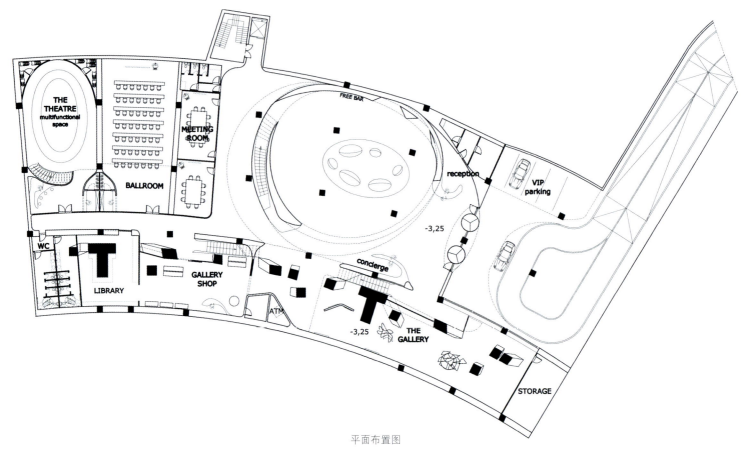

平面布置图

Bird Nest Art Hotel

With the completion of the iconic Beijing Bird Nest Stadium, many design enthusiasts have been anticipating the next moves of this unique building. Since the short-range physical goals have been reached, the remote management target is assumed to become the next top priority. Beijing Bird Nest Art Hotel proposal, is definitely a combination of unique design and implemental long-term business strategy's best representative.

Using surrealism as the conceptual design input, it starts from Bird Nest Stadium itself. Imegin adding the commercial elements into this unique design stadium and

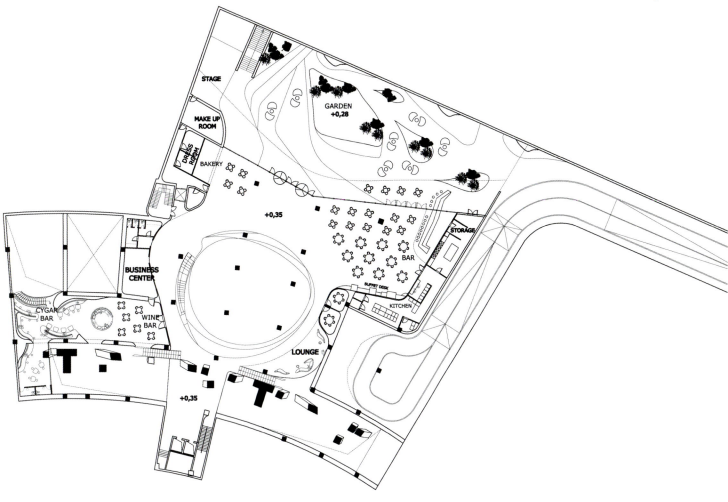

平面布置图

how to respond to its own as the integration of art and reality without losing the creativity which is the main challenge of our proposal. With the avant-garde stadium appearance, breaking away from the reality of the dream—like landscape with the interior echoes, it shows a strong visual highlight.

In our proposal, the design inspiration of Beijing Bird Nest Art Hotel emanates from the egg shape mixed with parametric design. Getting inspiration from Surrealism artists, especially Escher, we shifted the art theme of impossible realities to architecture. Using the latest design and parametric software, we developed the impossible surfaces starting from a particular equation of the Klein Bottle surface.

There is a wavy staircase in the lobby and some egg-shaped seats juxtaposed inside the space. The color tone is simple and clean – it is pure white and some tinge of purple. The shape of the staircase is so unique that people can even use this space as a fashion runway. The restaurant of Beijing Bird Nest Art Hotel also revolves around the theme of egg shape. There is plenty of curvy-shaped rattan furniture inside and a round-shaped bar counter at the front. There is a Japanese restaurant and a spa room which is also egg-themed. The apartment suites are also elegantly designed. There is a rotating wall in the living room and this wall is installed with LED technology. The vanity basin is placed in the open instead of being in the bathroom. The cascade of green leaves which drapes down the wall of the bathroom is also a design feature.

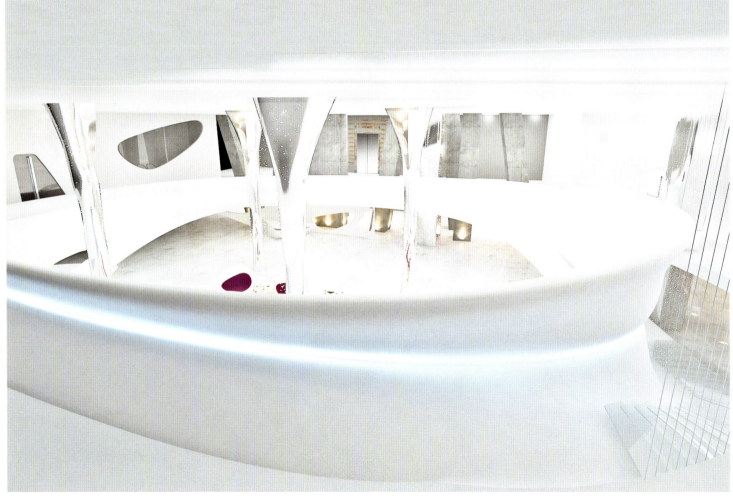

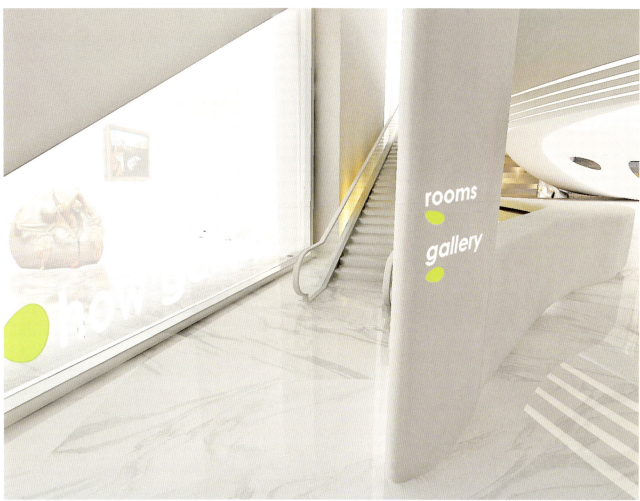

4

优胜奖 ／ Winning Prize

项目名称：泰和之春别墅样板房
设 计 者：陶金平
设计单位：苏州市建筑设计研究院有限责任公司

沉稳、洗练、环保打造出舒适低碳空间。

设计师走的是现代、简洁、创新的风格。色调沉稳、大方、时尚，线条流畅，平面布局合理是这套别墅的特色。同时还引用了新型的采暖空调形式——毛细管空调系统。它的优点是安装方便，高节能性能、无环境污染，使整个别墅智能、环保、节能化，增加了居住空间的舒适性。

客厅采用镀钛不锈钢"格子"结构，强调垂直线条，烘托空间挑高达6m的开敞格局，让人感觉大气奢华。灯光的点缀使得原本利落的空间又增加了一丝飘逸灵动，打造出豪宅的不凡气势。

餐厅延续了客厅的"格子"元素，使之更为统一和谐，精致的不锈钢酒架既用做隔断，又让用餐空间更为宽敞，充分考虑了视觉效果及实用性。

主卧套房讲究舒适及完善的功能。房间与主卫之间通过透明玻璃相连，床背景的茶色烤漆玻璃一直延伸到卫生间，使空间协调的同时又显得更加宽敞。吊顶的蓝色LED灯光犹如幽蓝的月夜般营造出浪漫的气氛。

设计师注重细节及品质的塑造，彰显空间整体氛围的和谐，结合高科技的运用更是相得益彰。步入各功能区，就像穿梭在沉静华美的世界。色彩、材质以及统一步调的渐进过度、交错，设计师用心缔造了一处优雅、低调、奢华的低碳、节能、舒适的居所。

Being sedate, concise and environmental protective creates the comfortable low carbon space.

The designers take the modern, concise and innovative style. The sedate tone, generous, fashion and fluent line and reasonable plane layout are the characteristics of this villa. Meanwhile draw into the new style-capillary air conditioner system. Its advantages are easy-being installed, high energy-saving performance and no environmental pollution, which makes the whole villa intelligent, environmental protective and energy conservation ,increasing the comfort of the living space.

Use the titanium coated stainless steel "grid" structure in the sitting room, and emphasize the vertical line to foil the open pattern of 6cm height and make people feel generous and luxurious. The ornament of lights makes the originally agile space more elegant and smart and creates the extraordinary momentum of the villa.

Continue to use the grid element of the sitting room in the restaurant, which make it more unified harmoniously. The delicate stainless steel wine frame is used as a partition which makes the dining space more capacious fully considering the visual effect and practical use.

The main bedding suite is particular about the comfort and perfect function. Use the transparent glass to connect the room with the main washing room. The dark brown baking varnish glass of the bed's background extends to the toilet and makes the space coordinate with each other and also makes the

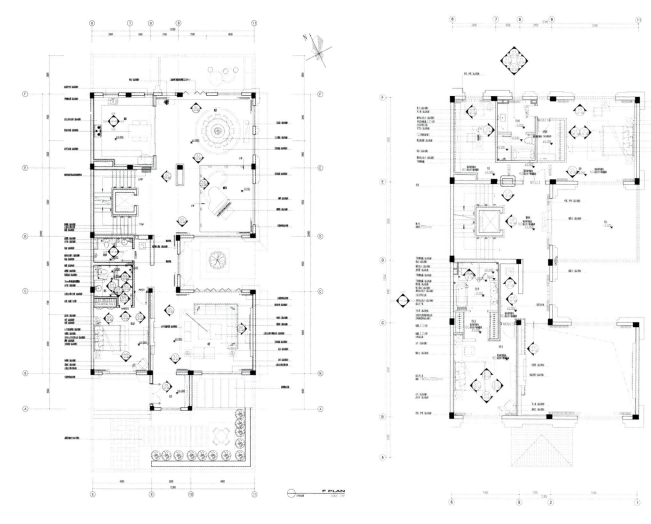

一层平面布置图　　　　　　　　二层平面布置图

space more capacious. The Blue LED lighting on the ceiling creates romantic atmosphere like the moonlit night.

The designers pay attention to the detail and quality, revealng the harmony of the whole space and combine with the high-tech which plays double roles. Walking into the function area is like shuttling in the calm and colorful world. The color, materials and the unification step gradually developing and crossing creates one elegant and low-key, low carbon, energy-saving and comfort living place with heart.

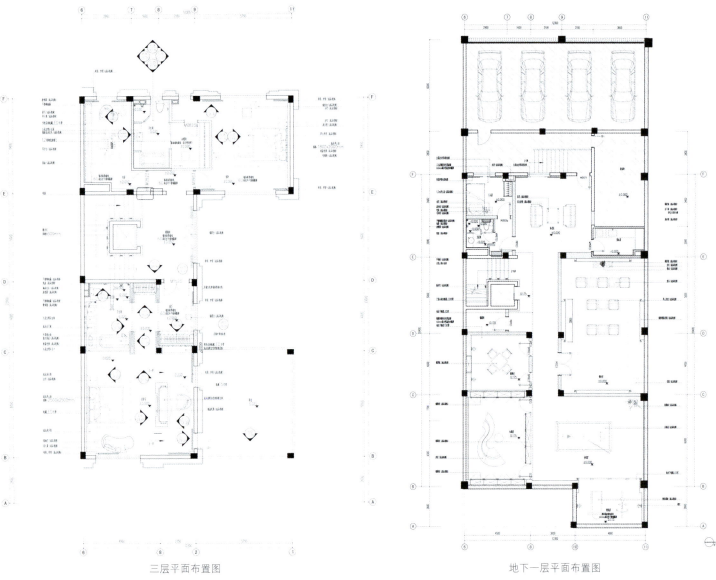

三层平面布置图   地下一层平面布置图

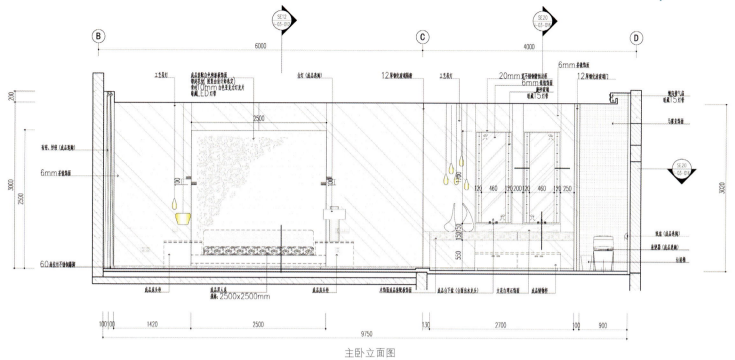

主卧立面图

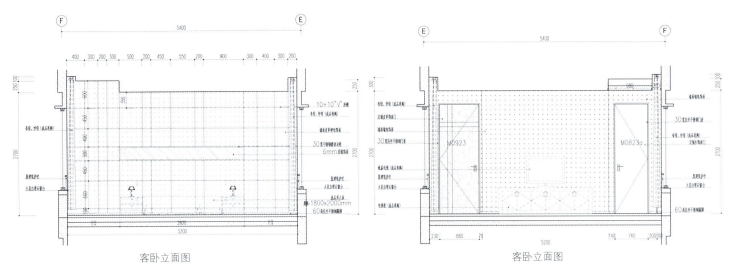

客卧立面图　　　　　　　　　　　　　　　　　　　客卧立面图

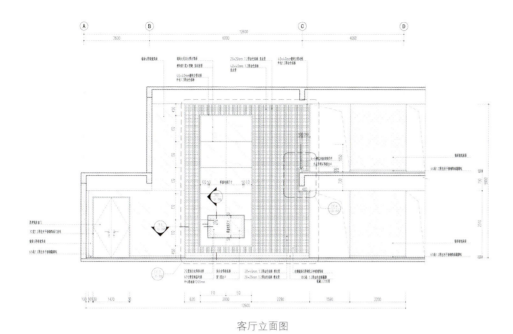

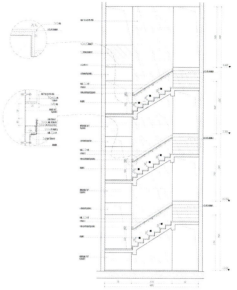

客厅立面图

楼梯立面图

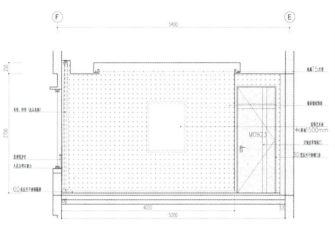

客卧立面图

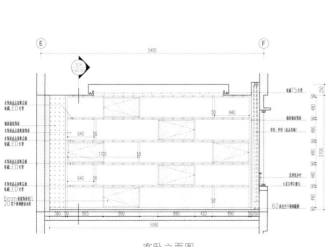

客卧立面图

项目名称：天津智谛山售楼处
设 计 者：张驰
设计单位：苏州建筑设计研究院限责任公司

天津滨海置地·智谛山售楼处改造设计

天津滨海置地开发的"智谛山"住宅项目位于天津塘沽经济开发区。售楼处从建筑到室内体现了年青人积极向上、活波的性格及时尚前卫的生活方式，同时还体现绿色环保的概念。由于该项目是由上世纪70年代老厂房改建而成，并且又地处冬季比较寒冷的北方，所以如何把新的视觉感受与老建筑相结合，又不破坏建筑基础，较好地保证节能环保效果，是设计师面临的巨大挑战。

设计师从如下三个方面入手解决问题

1.首先为老厂房穿上新衣。厂房表面加装穿孔铝板、彩色氟碳喷涂，不仅在严寒冬季缺少绿化的北方，为老建筑带来活泼新鲜的视觉感受，还在夏季能过滤多余的阳光和保证透气通风，使室内保持凉爽。在铝板上又加装了中空断桥玻璃幕墙，出色的保温效果在寒冷的冬季得到了充分的肯定。

2.其次，在室内运用了大量天然环保材料，如天然木地板、烧毛石材，特别是一排排金属格栅上做的绿色植物喷绘，让人仿佛进入了郁郁葱葱的竹林，一派绿色生机、春意盎然的景象，让人精神一振。洽谈区做了一面水墙，水流如丝缎般在天然毛石表面滑过，流入水槽，与室外水槽流通。在炎热的夏季不仅调节了室温，还在干燥的冬季起到了调节室内湿度的作用。

3.最后采用了一些新型节能技术，如安装了日光采集照明，白天它能将室外光线采集到室内，而室内不必再使用任何可用电照明，节省了能源；在外墙安装了保温层（双层玻璃幕墙），保证了冬季室内的温度；在样板房展示区开辟了内庭院，保证样板房及售楼处的自然通风；这些利用天然能源的措施使本售楼处不仅靓丽美观，而且带给使用者较高的舒适度，同时也体现了节能环保、绿色的理念。

Tianjin Bin Hai Land·Zhi Di Mountain Sales Office Reconstruct Design

"Zhi Di Mountain" Housing Project developed by Tianjin Bin Hai Land Co., Ltd is located at the Tang Gu Economic Developing Zone in Tianjin. The design of the sales office shows the young's positive and active characters and fashion and avant-garde life way from building to indoor. Because the project is reconstructed from the old factory in 1970s and is in the north where it is very cold in the winter, it is the huge challenge for the designers that how to combine the new visual experience and old buildings but not to destroy the building foundation and guarantee the energy conservation and environmental protection.

The designers solve the above problem from the following three aspects:

1. Put the new clothes on the old factory workshop. Set the perforated aluminum plate on the workshop surface and paint the colorful fluorine carbon, which not only brings the active and fresh visual experience to the old buildings in the north where it is cold in the winter and short of green and filter the redundant sunshine ,ensure the ventilation and keep the indoor cool. The added insulating and bridge-cut glass curtain wall and the excellent heat preservation effect in the seriously cold winter get affirmed fully.

2. Use the lots of natural environmental protection materials indoor, such as the natural wood floor, singeing stone, especially the rows of green plant paint on the mental grille make people feel walking into the green and luxuriant bamboo forest and the green vitality makes people have springs and moods.

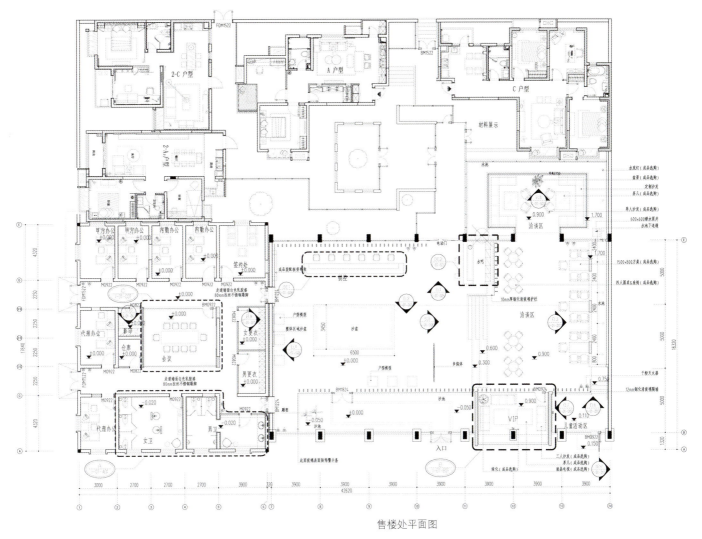

售楼处平面图

A wall is built in the negotiating area and the water flows through the surface of the natural rubble like silk into the water trough and connects with the trough outside, which not only adjusts the indaor temperature in the hot summer but also adjusts the humidity in the dry winter.

3. Adopt some new energy-saving technologies, such as fixing daylight harvesting lighting, during the day, it can harvest the outdoor light into indoor and there is no need to use any electric lighting which saves the energy. Fix the heat-preservation layer on the exterior wall (double layer glass curtain wall), which keeps the temperature in the winter. Making a courtyard in the model room exhibition area keeps the natural ventilation of the model room and sales office. The measures using the natural energy not only makes the sales office beautiful but also brings the better comfort to the users and develops the energy conservation, environmental protection and green concept.

售楼处顶面图

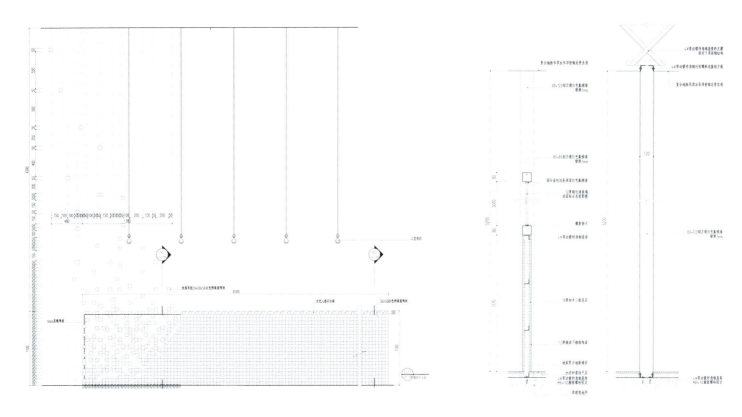

销售台大样图   大厅立面图

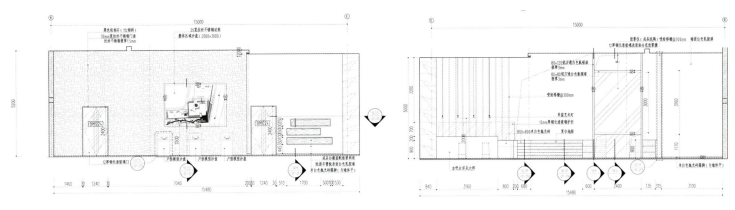

大厅立面图   大厅立面图

项目名称：福州世欧地产王庄售楼处及会所
设 计 者：田饶、林玲（室内事业部深圳区域团队）
设计单位：中建国际(深圳)设计顾问有限公司

会所【世欧地产王庄一期地块会所】

会所位于整个楼盘中心地带的下沉式广场的中央位置，以"诺亚方舟"命名，位置显耀。

会所的环境和室内设计以水为主题，会所周边围绕的水景，使"方舟"得以舟行水上，而水中花则成为室内设计的主题，会所的平面和立面都极富流动感，水与花相结合，将观者引入美轮美奂的会所空间。

会所由养生茶艺区、咖啡区、网上冲浪区和儿童活动区组成，平面功能划分明确而灵活，隔墙似花团锦簇，与建筑设计相得益彰，开放式的功能分区使来宾感觉轻松自在。

由观光电梯下到会所大堂，步入明亮的大厅，映入眼帘的是白色弧形的墙面、红色的花形座椅，墙上的细碎花朵造型线条细腻，在柔和灯光下显得格外生动，所有这一切，演绎着别致的现代风格。

养生茶艺区分为开放式休闲区与私属会客室，略带中式设计风格的私属会客室用花瓣形的木制高隔断及栅格状的半隔断围合起来，品茶之余，营造静谧的休闲氛围。

休闲咖啡区有丰富的娱乐休闲设施，如沙弧球、台球、飞镖，令氛围更加生动活跃；以弧形书架围合出的休闲书吧让人在活动之余可以安逸地窝在沙发里，随手翻阅自己心爱的书籍；透过咖啡厅的玻璃可以欣赏到室外美景，绿色植物、瀑布的美景与室内的轻松氛围融合，令人轻松自在，心旷神怡。

在咖啡厅的一角还设有红酒雪茄房，红酒雪茄房隐蔽而不偏僻，是一个高端的休闲空间。现代奢华的设计，为来宾提供品酒品烟、商务会谈，安静而又舒适的场所。

同时儿童游乐区让孩子们也能在此找到乐趣，绘画班、DIY室以及多媒体房为少年儿童提供娱乐场所，寓教于乐。

总之，会所为人们提供了一个休闲娱乐的好去处。

Club Phase I Wangzhuang Plot Club of Show Real Estate

Situated at the center of sunken square in the whole estate's heartland, the club is named after Noah's Ark with a renowned location.

Environment and interior design of the club adopt water as theme. The surrounding waterscape makes Noah's Ark look like sailing on the water, and flower in water appears to be motif of interior design. All the seemed flowing planes and elevations of the club and combination of water and flower provide splendid club space to viewers.

The club consists of health tea space, coffee space, internet surfing space and children playing space. With blossom-like partition, plane is divided for functions clearly and flexibly, complementing architectural design, which makes visitors feel at ease.

Get down to lobby by sightseeing lift, step into the bright hall, and then you would see the white arc-shaped wall finishing, as well as red flower-shaped chairs. The refined small broken beautiful flavers on the wall seems extraordinarily vivid in mild lighting. All these things are to be interpret a chic modern style.

Health tea space is divided into opening lounge and private parlors which possess Chinese design style to some extent and are enclosed by petal-shaped high wooden partition and lattice half-partition together. Visitors could enjoy the tea as well as quiet and leisure atmosphere at the same time.

There are plenty recreation facilities in the leisure coffee space, such as shuffle-board, billiards, and dart, which make the atmosphere more lively and active; the leisure book bar encircled by arc book shelves enables people to read favorate books in sofa cozily after activities; in this space you could enjoy outdoor scenery through the glass of coffec shop, and the integration of plants and waterfalls and relaxing atmosphere indoor makes visitors carefree and refreshed.

In a corner of the coffee space, there is a wine-and-cigar room, which is a concealed but not remote high-end leisure space. The extravagant design,serves as a quiet and comfortable place for visitors to taste wine and tobacco and have business meeting.

Children could have fun in children playing space, where there are painting classes, DIY rooms and multi-media rooms to provide entertainment place for children learning through playing.

In a word, the club is an admirable place for leisure and recreation.

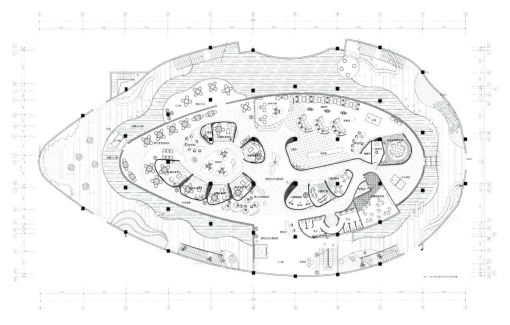

地下层平面布置图

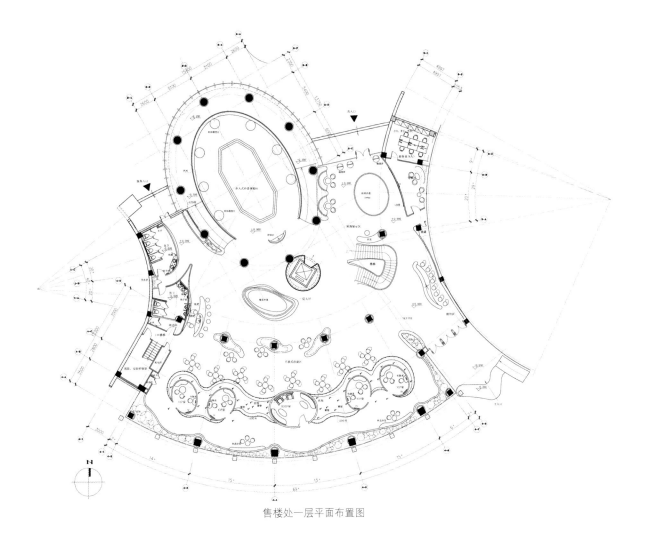

售楼处一层平面布置图

售楼处【世欧地产王庄售楼处】

福州世欧地产王庄楼盘位于福州市旧城区的黄金地块，周边环境热闹繁华；售楼处是一期地块社区规划中的幼儿园，设计师从建筑的外廓中提取层层叠叠的圆形元素，结合福州当地靠海临江的地理人文，从中萌发出"涟漪"的设计理念。

水中涟漪，心中涟漪，陶醉于美轮美奂的空间臆想之中，结合设计理念，将VIP室置于钢制荷花光影水池中，通透的玻璃幕墙将全景映入观者眼帘，水中涟漪的感觉透过另类的表达方式淋漓尽致地体现出来。

设计中倡导运用一系列的环保材料，主要包括白、灰、黑人造石，复合木塑板，自然采光最大化的屋顶，可渗透景观等。进入建筑内，映入眼帘的是群聚的异型人造石柱体、曲线特异的隔墙造型及家具，移步异景，无不突显现代国际居住社区的氛围，同时达到灵动美观的空间意境。

Sales office Wangzhuang Sales Office of Show Real Estate

With bustling surroundings, Fuzhou Wangzhuang estate of Show Real Estate lies at golden area of old city proper in Fuzhou; asles office is aperiod in the planming of the kindergarden plot community. Extracting layers of circles from profile outside building, syncretizing seaside and riverside culture, designer sprouts the design idea of "ripple".

Watching ripple in water, you may feel ripple in heart, and intoxicated with magnificent space imagination. The VIP room is designed in steel pool where lotus chequers with light and shade. Panoramic view greets viewers through transparent glass curtain wall, showing the ripple vividly by am unique way.

A series of environment-friendly materials are proposed, mainly including rostone of white, grey and black, composite wood-plastic plate, roof available for maximized natural light and permeable landscape. Upon entering the building, you could see the gather of special-shaped rostone pillars as well as partitions and furnitures of specific curve. Each movement of step will bring the different sceneries. All these present atmosphere of modern international residential community, and meanwhile a livingly and beautiful space.

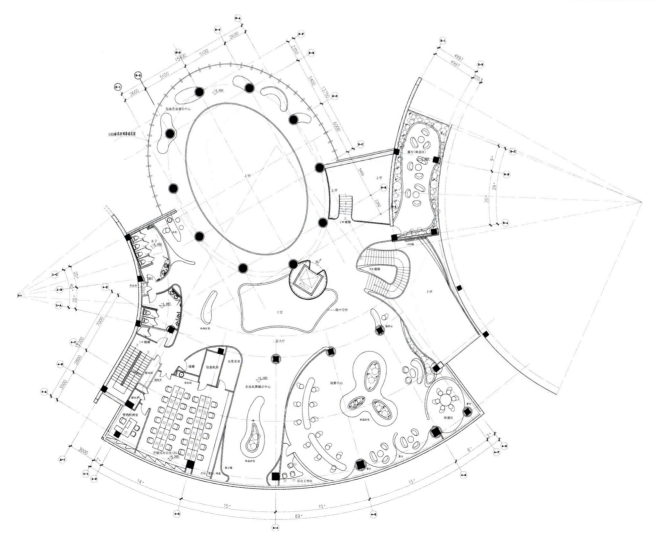

售楼处二层平面布置图

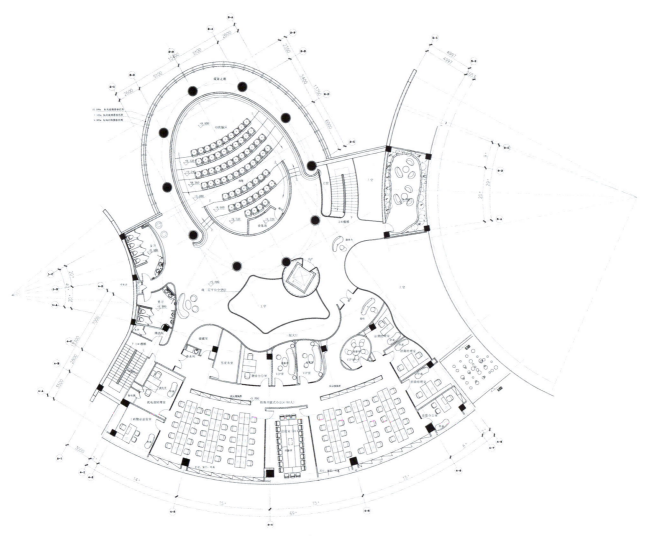

售楼处三层平面布置图

分　　类：3Quah Paik Choo Bernadette
　　　　　(Bernie)(PPK Kindergarten)
风　　格：真材实料
概　　念：循环思维
大　　学：KBU INTERNATIONAL COLLEGE KBU国际大学
讲　　师：MS.JOANNE LOW
名　　字：QUAH PAIK CHOO BERNADETTE

PPK幼儿园

重新定义可持续性的不费力的学习方式。

PPK提供了一种多感官的可持续性的学习方式。好的学习风气可以促进、启发孩子的潜能。

3C综合楼

四层高的综合楼，作为一个儿童和年轻人的开发中心，由MPSJ市政委员会的Subang Jaya所设计，以适应快速增长的人口。

地下层是PPK幼儿园，用于迎合MPSJ员工的孩子需要（3～4岁和5～6岁）。此项目为幼儿园的水平提升提出方案。

循环思维

空间设计受启发于大片的外部空间（天空现象）和现有场地的内部空间，并且有潜力更好地利用这块地方。

循环思维的概念突出了思维过程和孩子与学习环境的互动，以此作为教育的刺激方式。

材料还原本色

目的是减少材料的表面处理，并且尊重其自然特性。裸色混凝土墙、再利用的木料和当地的竹材是此设计的主要使用材料。

教学法

环境、家长、老师和儿童的互动教学法。

01 表面设计

垂直的地下通风通道，热空气上升离开通道，利用大气压从底部吸进冷空气。如果外界空气温度平均分布，上部的风吹进通道迫使热空气下沉。

实际情况：两种过程同时进行。无论哪种方式，通风管道总是促进空气流动。此建筑朝向高速路，面对空气污染。建筑表面设计回应了这个问题。外突的玻璃墙是噪音和大风的阻隔，保护着竹子环绕的人行道。竹子的净化能力净化了污染的空气，提供室内清新的空气。

颜色寻路：从幼儿园入口处到内部颜色的渐变告诉孩子他或她在什么地方。

绿色：欢迎的颜色——社区空间，家长可在此逗留并参加一些儿童活动。

黄色：愉悦的颜色——可刺激人的心理，更好地吸收教授的知识和传达的信息。

黄色：营造活跃的氛围，在孩子们之间增加互动，培养良好的社交技能。

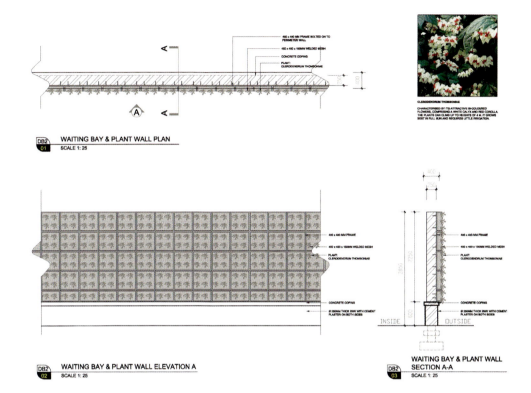

CLERODENDRUM THOMSONIAE
CHARACTERISED BY ITS ATTRACTIVE BI-COLOURED FLOWERS, COMPRISING A WHITE CALYX AND RED COROLLA. THE PLANTS CAN CLIMB UP TO HEIGHTS OF 4 M. IT GROWS BEST IN FULL SUN AND REQUIRES LITTLE IRRIGATION.

WAITING BAY & PLANT WALL PLAN
SCALE 1: 25

WAITING BAY & PLANT WALL ELEVATION A
SCALE 1: 25

WAITING BAY & PLANT WALL SECTION A-A
SCALE 1: 25

植物墙体：3C 建筑周围的植物墙既提供了让人欢欣的美观的效果，又阻挡了附近公路的尾气和噪声污染。

水力发电水景：在花园里安装了一个水力发电机，利用水流产生的动能为幼儿园发电。PPK的设计注重美观。

纤维管道空调系统：整个幼儿园使用了纤维管道空调系统，使得冷气均匀分布，减少了普通空调存在的制冷死角。

02　等候区

接乘区域。从LG层通过楼梯直接进入幼儿园。

03　中庭

中庭的设计为了促进儿童、教师和家长间的互动。

这是一个宽大的社区空间，从入口处开始。在这里家长可以驻足甚至参加一些儿童活动。这是一个休闲的区域，可以直接看到学习空间，并在竹制雕塑上展示孩子们的作品。

原型：竹子雕塑展示系统

04　教师休息区

宽敞的员工空间包括会议室、办公室、休息室以及提供午餐的厨房。环境的绿色基调为老师提供了安静的氛围，以便更好地休息。

05　儿童洗手间

利用来自天井的自然照明的设计思路。

浴室的水源支持旁边的植物生长，同时植物也使室内充满香味。

06 教室

多色天花板灯完全由外界的能源提供，供电无需使用电线和电池，利用发光涂料清除周围像CFL和LED灯泡产生的能源废物。无论白天还是晚上都会持续提供一种柔和的环境光源。

此空间设计为人群活动及休憩场所。有专门为每个儿童个人物品的存放处，还有展示项目及手工的场所。洗手间方便到达。

装具材料包括再生材料和当地供货商提供的产品，让人感到强烈的当地社会、经济和生态系统的氛围，比如说，使用当地的木料和竹子。

素馨色是一种让人愉悦的颜色，可刺激人的心理，更好吸收教授的知识和传达的信息。

07 普通区域

此处为所有教室的主要循环空间，协助孩子找到通道，提供孩子们更多交流的机会。

08 普通区域

学校利用此空间进行特殊活动，例如音乐会和戏剧演出。

橘红色营造一种活泼的氛围，促进孩子们的互动交流来开发社交技能。

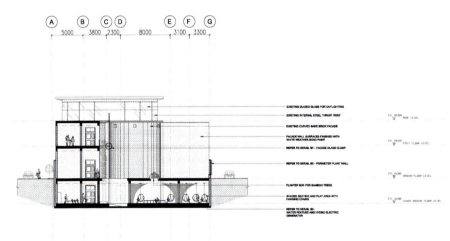

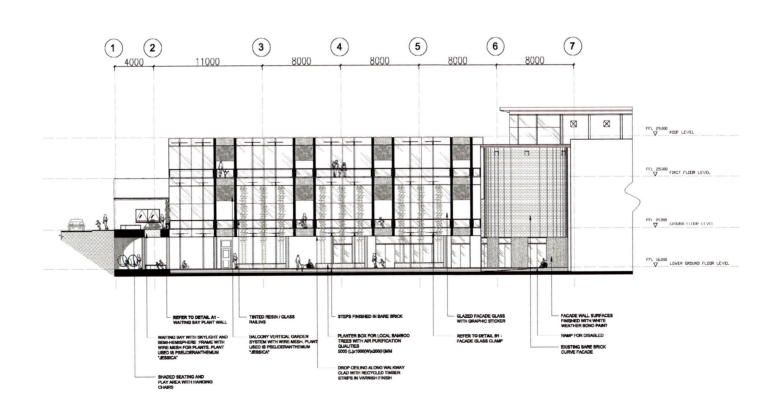

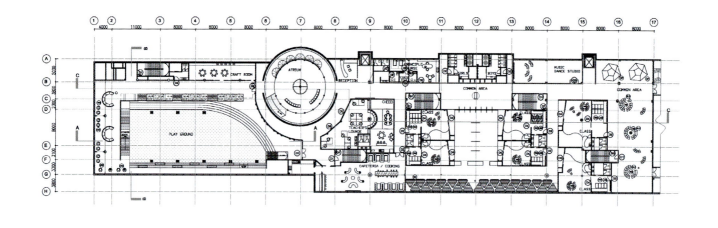

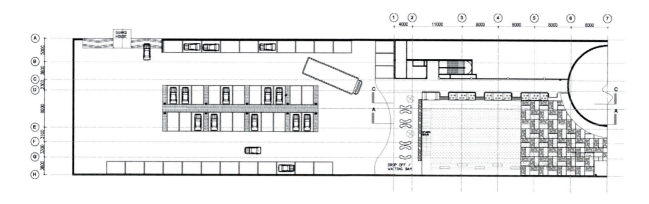

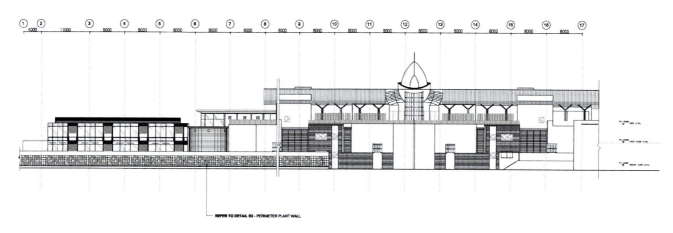

REFER TO DETAIL 82 - PERIMETER PLANT WALL

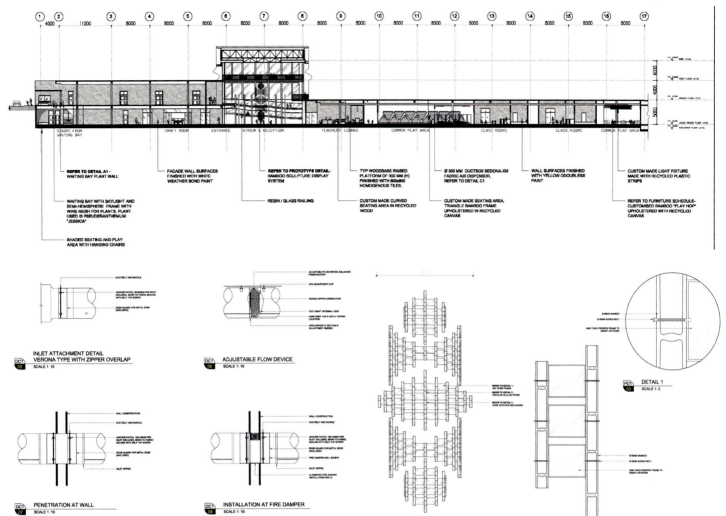

SUSTAINABLE – COMFORTABLE SPACE

2011 ASIA PACIFIC INTERIOR DESIGN AWARDS FOR ELITE

PROJECT BRIEF

PPK Kindergarten

Learning redefined. Effortless sustainability

PPK offers a multi-sensory sustainable way of learning. The ethos is choreographed around individual moments that stimulate and inspire. This project is a proposal for the upgrading of the existing kindergarten in Subang Jaya that is in need of renovation.

CONCEPT

The concept "Cycle of thought" was inspired by the open outdoor and indoor space available of the existing site and its potential to be better utilised. The concept highlights the thought process and spatial interaction between a child and the learning environment as a stimulating way of education. The space is designed to encourage pedagogy of interaction between environment, parents, teacher and kid

STYLE

Truth in Materials

The aim was to reduce the use of material surface treatments and celebrate its natural character.

Bare concrete walls, recycled timber and local bamboo are the main materials used in the design.

SUSTAINABILITY

Climate, Energy and Water

Hydroelectric Water Feature

Reduce water in bathroom

Celebrate water...a luxurious shower bridges indoors and out. Water used for showers also maintains the adjacent herbs—that in turn scent the room.

Building resources & Ecology Environmental - Materials

The furnishings embrace repurposed materials and local vendors, providing a strong connection to the local community, economy, and ecosystem.

Bamboo Facade

Recycled timber ceiling finish

Recycled canvas (common area-tunnel)

Recycled cardboard roll shelving (teachers' lounge)

To reduce the noise level in the cafe area, the carpet feature

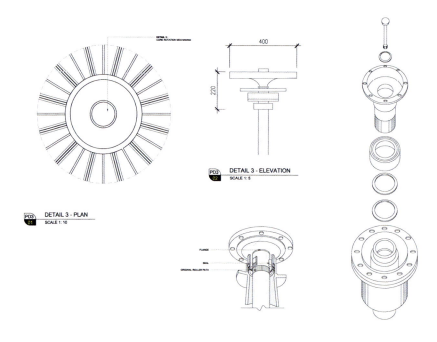

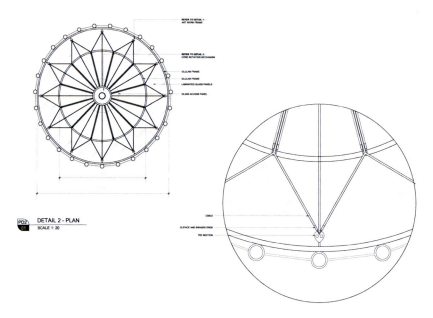

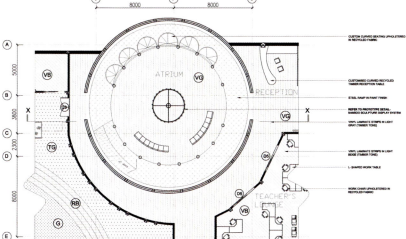

wall absorbs sound to create a soothing environment.

Rattan Mats (Class rooms)

Luxurious natural bedding and materials as well as plentiful natural ventilation offer restful sleep.

Enhance Indoor Environment Quality (IEQ)

Day lighting

Ventilation – fabric air duct

To increase the efficiency of the AC system, Air dispenser fabric ducts are installed. It creates an even distribution of cool air, reducing cold spots like common AC outlets.

.Low VOC Paints

The colour scheme – Green, Yellow, and Orange - is used throughout the design as a method of way-finding. The progression of colour and forms informs the child where he or she is.

Green innovation - Operational and Maintenance Practices

Hydro Turbine

The kinetic energy of the water flow is able to create electricity to be used for the store by installing a small Hydroelectric

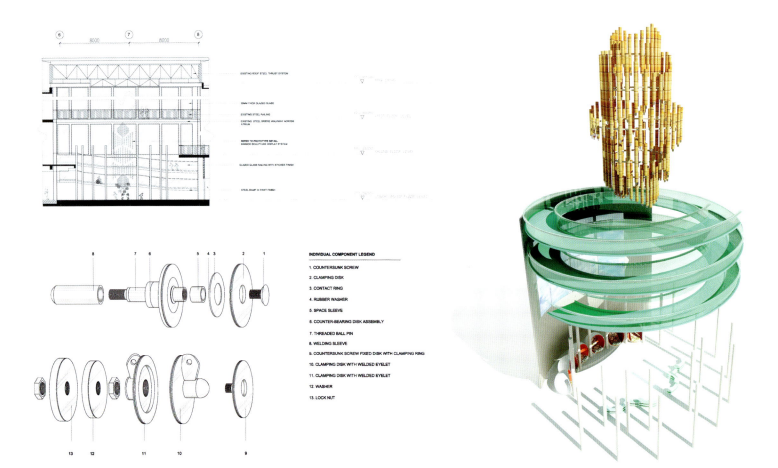

INDIVIDUAL COMPONENT LEGEND

1. COUNTERSUNK SCREW
2. CLAMPING DISK
3. CONTACT RING
4. RUBBER WASHER
5. SPACE SLEEVE
6. COUNTER-BEARING DISK ASSEMBLY
7. THREADED BALL PIN
8. WELDING SLEEVE
9. COUNTERSUNK SCREW FIXED DISK WITH CLAMPING RING
10. CLAMPING DISK WITH WELDED EYELET
11. CLAMPING DISK WITH WELDED EYELET
12. WASHER
13. LOCK NUT

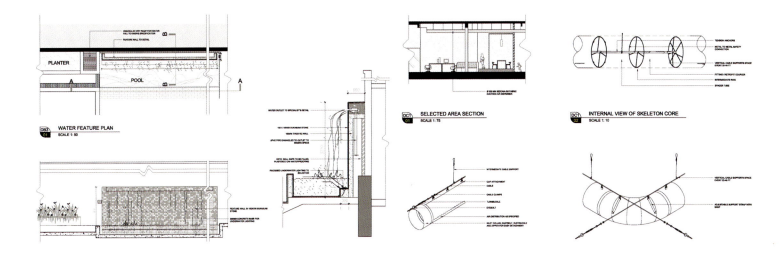

Generator water feature. Decorum's design aims for beauty with a purpose.

Green wall / shade

Façade

Facing the express way, the building is exposed to pollution. The facade design responses to this problem. An extruded glass wall acts as a noise and wind barrier protecting the walkways surrounded by bamboo. The purification qualities of bamboo purify the polluted air and provide clean air to the interiors.

Vertical gardening provides an eco and aesthetic look.

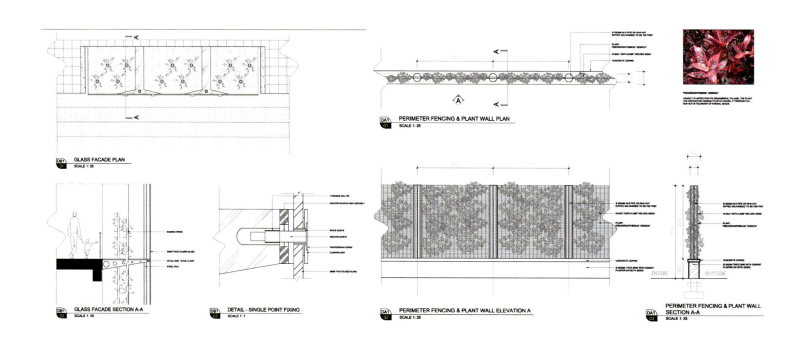

GLASS FACADE PLAN
DB1 01
SCALE 1: 25

PERIMETER FENCING & PLANT WALL PLAN
DA1 01
SCALE 1: 25

GLASS FACADE SECTION A-A
DB1 02
SCALE 1: 25

DETAIL - SINGLE POINT FIXING
DB1 03
SCALE 1: 1

PERIMETER FENCING & PLANT WALL ELEVATION A
DA1 02
SCALE 1: 25

PERIMETER FENCING & PLANT WALL
SECTION A-A
DA1 03
SCALE 1: 25

图书在版编目（ＣＩＰ）数据

可持续舒适空间：亚太室内设计精英邀请赛获奖作
品精选集. 下 / 亚太设计中心编著. -- 南京 ：江苏人民
出版社，2012.7
　ISBN 978-7-214-08328-9

　Ⅰ．①可… Ⅱ．①亚… Ⅲ．①室内设计－作品集－亚
太地区－现代 Ⅳ．①TU238

中国版本图书馆CIP数据核字(2012)第120127号

可持续舒适空间：亚太室内设计精英邀请赛获奖作品精选集（下） 亚太设计中心 编著

策划编辑：刘晓华
责任编辑：蒋卫国　张 蕊
责任监印：彭李君
出版发行：凤凰出版传媒集团
　　　　　凤凰出版传媒股份有限公司
　　　　　江苏人民出版社
　　　　　天津凤凰空间文化传媒有限公司
销售电话：022-87893668
网　　址：http://www.ifengspace.cn
集团地址：凤凰出版传媒集团（南京湖南路 1 号 A 楼 邮编：210009）
经　销：全国新华书店
印　刷：深圳当纳利印刷有限公司
开　本：965 毫米 ×1270 毫米　1/16
印　张：17
字　数：136 千字
版　次：2012 年 7 月第 1 版
印　次：2012 年 7 月第 1 次印刷
书　号：ISBN 978-7-214-08328-9
定　价：258 元（USD 48）
　　　（本书若有印装质量问题，请向发行公司调换）